Praise for
THE ZOMBIE AUTOPSIES

"A chilling, and thrilling, look into the anatomical makeup of those brain-gobbling, reanimated freaks. Awesome."
—*Boston Phoenix*

"[A] clever debut...Schlozman makes the science both accessible and plausible...This book is sure to be scarfed up by ravenous zombiephiles." —*Publishers Weekly*

"Dr. Steve's THE ZOMBIE AUTOPSIES will charm and excite a new generation into loving science."
—Chuck Palahniuk, *New York Times* bestselling author

"Entertaining...Schlozman weaves a frightening scenario, and horror fans will admire illustrator Sparacio's grisly drawings of the disease's progress...nicely complements Max Brooks's *The Zombie Survival Guide*." —*Kirkus Reviews*

"An inventively framed apocalyptic tale embellished with black humor." —*Booklist*

"Gruesome and gripping! Steven Schlozman reveals the science behind zombies from the inside out."
—Seth Grahame-Smith, *New York Times* bestselling author of *Abraham Lincoln: Vampire Hunter*

THE
ZOMBIE
AUTOPSIES

THE
ZOMBIE
AUTOPSIES

Secret Notebooks from the Apocalypse

BY STEVEN C. SCHLOZMAN, MD

Illustrations by Andrea Sparacio

GRAND CENTRAL
PUBLISHING

NEW YORK   BOSTON

Copyright © 2011 by Steven C. Schlozman, MD
"There Have Been Questions..." Copyright © 2012
by Steven C. Schlozman, MD

Grand Central Publishing
Hachette Book Group
237 Park Avenue
New York, NY 10017

www.HachetteBookGroup.com

Printed in the United States of America

Originally published in hardcover by Grand Central Publishing.

First Trade Edition: March 2012

10 9 8 7 6 5 4 3 2 1

Grand Central Publishing is a division of
Hachette Book Group, Inc.

The Grand Central Publishing name and logo is a trademark of
Hachette Book Group, Inc.

The Hachette Speakers Bureau provides a wide range of
authors for speaking events. To find out more, go to
www.hachettespeakersbureau.com or call (866) 376-6591.

The publisher is not responsible for websites (or their content)
that are not owned by the publisher

The Library of Congress has cataloged the hardcover edition
as follows:
Schlozman, Steven C.
The zombie autopsies : secret notebooks from the apocalypse /
Steven C. Schlozman. — 1st ed.
p. cm.
ISBN 978-0-446-56466-3
1. Zombies—Fiction. I. Title.
PS3619.C435Z66 2011
813'.6—dc22
2010028950

ISBN 978-0-446-56465-6 (pbk.)

To Ruta, Sofia, and Naomi,
for putting up with me.

THE
ZOMBIE
AUTOPSIES

UNITED NATIONS OUTPOST

Highly Confidential Memo

25 January 2013

The enclosed documents are highly classified. They are exact replicas of the recently recovered handwritten notes of Dr. Stanley Blum, the last scientist sent to the United Nations Sanctuary for the study of ANSD and "zombie" biology. After careful examination, World Health Organization and United Nations officials have reason to believe that these writings may contain groundbreaking data regarding the nature of the ANSD pathogen. Furthermore, it is possible that this information represents both a key to a viable cure, as well as the most definitive evidence to date that the ANSD virus was artificially manufactured—**human made**—and therefore did *not* occur naturally. The importance of these potential conclusions cannot be overstated.

One-third of humanity has perished from the plague. Two point three billion people have died, and countless more are quickly moving toward the final stages of disease. There is reason to believe that in a short time nearly everyone on Earth will be infected. The virus continues to spread exponentially, and all attempts at a vaccine or a cure have failed. Scientific and industrial infrastructure is rapidly faltering. Early attempts at controlling the spread of disease via nuclear and non-nuclear incineration have left the planet in an ecologically fragile state. Current computer models suggest that civilization can only survive for approximately another decade before we face total destruction. These are indeed dire times. The information in this journal may very well represent our last hope.[1]

BACKGROUND

<u>Less than three years ago, the idea that a "zombie plague" would threaten all of humanity was preposterous. Indeed, the very nature of this claim prevented swift and decisive action from the world community. Nevertheless, "zombie" is arguably the most recognizable term for those who have been infected with ANSD and have reached the fourth stage of the disease.</u>

Ataxic Neurodegenerative Satiety Deficiency Syndrome (ANSD), the internationally accepted diagnostic term for what is more commonly referred to as zombiism,

---

1. In addition to new material, the following documents contain background information intended to brief the fourteen new representatives who have recently arrived at the UN compound. The WHO and the UN regret the contamination that necessitated the removal of those with Stage IV infection in last month's breach. Precautions have been taken to prevent further infectious incidents. Early identification protocols remain in effect.

continues to spread unabated. While some islands remain disease free, ANSD is otherwise present on every major landmass. Most of the governed world is under martial law. The signs and symptoms of the disease are universally known, but the disease itself appears both indestructible and untreatable.

In July 2012, the United Nations established a study site on the island of Bassas da India in the Indian Ocean. The official charter for this site called for a laboratory setting where the world community could focus its efforts on the scientific study of ANSD, including anatomic explorations of zombies themselves, as well as molecular investigations of the presumed contagion. This site is formally known as the United Nations Sanctuary and Study Site (UNSaSS) but is more commonly referred to at the UN bunkers as the Crypt. Those who volunteer to participate in the work on the island understand that a return to the bunkers is not permitted and scientists started referring to the site as the Crypt in an effort to acknowledge the blunt reality of their important and difficult tasks.

On 11 November 2012, the UN received its last message from the Crypt. The message was a radio transmission, and, as with most digital communications, the content was badly garbled due to nuclear interference. It is important to note that we had not received any e-mail contact from the island for some time, and the voice message was received over a little-used commercial satellite frequency. Voice analysis has suggested with reasonable certainty that the communication was from Dr. Blanca Gutierrez, who was at that time the resident microbiologist on the island. After significant digital analysis, the message reads as follows:

Status...gent. Hype........new. Diff...virus.
Vaccine...ble

An international group of scientists, policy makers, epidemiologists, and ethicists has theorized that Gutierrez had new information regarding a viable vaccine. It is unclear whether "hype" refers to "hyper" (a possible acceleration of disease progression) or, more likely, to the hypothalamus, a region of the brain oddly not affected by ANSD infection. The "-ble" suffix at the end is consistent with words such as "possible," "viable," or, conversely, "impossible." We are fairly certain that "gent" is most consistent with the word "urgent."

No further communications were received, and attempts to communicate with the island were unsuccessful. At that time, UN records show that Gutierrez's team included Dr. James Pittman, a Canadian anatomist and medical illustrator; Dr. Anita Gupta, a leading virologist from Delhi; as well as three military attachés. There were also an unspecified number of infected humanoids in the holding facility.

On 14 November 2012, a team of three individuals was sent to investigate the Crypt in an attempt to regain contact with Gutierrez and her colleagues. Dr. Sarah Johnson, a Scottish neurobiologist and expert in viral brain infections; Dr. Jose Martinez, the chief forensic pathologist for the former New York City; and Dr. Stanley Blum,[2] a neurodevelopmental biologist with the United States Centers for Disease Control (CDC), were selected for the mission. While Drs. Martinez and Johnson were chosen for their scientific expertise, Blum's orders were to record the findings of the team and to ensure that these findings were reviewed by scientists in the international ANSD research community.

All three set out in separate automated transport planes programmed to bypass the security apparatus at the UNSaSS. Unfortunately, only Dr. Blum arrived safely on the island. After an automated transmission from Dr. Blum signaling his arrival, communication with the island was lost altogether.

At that time, ANSD experts felt the risk of further exploration of the Crypt was too great. Researchers on the island were genetically manipulating the virus, and ANSD experts were consequently reluctant to potentially release an even more dangerous form of the contagion from the sterile confines of the study site. However, as already

---

2. Many will recall that Dr. Blum was central in the research and diplomatic endeavors resulting in the seminal paper on ANSD published in August 2011. Prior to the completion of that manuscript, there was ongoing debate in the international community about the extent to which scientists understood or were prepared for the pandemic. Blum's diplomatic efforts brought together an international team of ANSD experts, succinctly described the history of the outbreak, and made clear the need for urgent attention to the rapidly developing crisis.

noted, world conditions continued to rapidly deteriorate. The decision was therefore made to pursue excavations of the Crypt with the primary goal of understanding the final transmission from Gutierrez.

After extensive precautions, the study site was thoroughly investigated for seven days, 8–15 January 2013. Three of the seven surveyors have since died from ANSD, and the remaining four have all shown signs of progressive infection. We are grateful for the courage and sacrifice made by those who conducted the surveillance.

Surveyor 3, Captain Amy McBride of the elite British Special Boat Service, was among the first to succumb to Stage IV disease while on the island. As protocol dictated, she was fatally shot by members of the surveillance team. Although she had collected only limited data onto her hard drive, the team failed to search her personal belongings, presumably in their efforts to quickly sterilize the body for transportation back to the UN base. She was frozen in liquid nitrogen and sealed in an automated marine transport unit. Navigational malfunctions prevented the remains of Captain McBride from arriving at the docking station until last week. Upon arrival, the handwritten laboratory notes of Dr. Stanley Blum were discovered in her backpack.

Dr. Blum's writings are presented here in preparation for tomorrow's meeting. His journal, as noted, is handwritten, and he and Dr. Pittman made extensive sketches and diagrams to document their work. The final pages of his notebook are frustratingly unclear, and physicians here speculate that he was traumatized as well as suffering from the cognitive effects of the treatments used to slow progression of ANSD toward zombiism.

The standard of care for treatment of ANSD infection remains an artificially induced increase in human pH. This technique was first described by Blanca Gutierrez and has been referred to as the Gutierrez protocol. Unfortunately, the effects of these measures create a number of significant neurological side effects that ironically mimic the cognitive decline consistent with full zombie status. Thus, it is unclear why Blum's cognitive state appears to decline throughout his notebook. He was either suffering from the effects of the Gutierrez protocol or was in fact slowly moving toward zombie status.

Nevertheless, it is possible that careful examination of this material will yield crucial answers. Most important, when Gutierrez left for the Crypt, she was already pursuing a theory that ANSD spread when it combined with still-unidentified additional infectious agents. If those agents could be detected, then there was a much greater likelihood of creating a viable cure. She therefore intended to pursue her theory by examination of infected humanoids while looking for additional pathogens.

These investigations had not yet been pursued by scientists at the Crypt or anywhere else, and Gutierrez insisted that she go herself. She felt strongly that investigations would have to be thorough and that findings from previous autopsies would need to be selectively discarded. Her expertise in molecular biology, microbiology, and vaccine development made her especially well suited for the position, though her commitment also meant that she would expire on the island. Thus, her efforts were both heroic and potentially scientifically groundbreaking. She asked that Dr. Anita Gupta join her as a co-investigator in this project, given Gupta's expertise in viral infections of the brain.

While Blum's journal documents Gutierrez's fate, we remain somewhat unsure how Blum himself expired. We are relatively confident that the decomposed body in the infirmary described by the surveyors was Blum's. Unfortunately, digital images of the site obtained by the surveyors are badly distorted by electromagnetic interference. Computer reconstruction teams are working to create interpretable images, but at the current time we have only the debriefings from the remaining surveyors, and none has the cognitive capacity to offer reasonable explanations.

The ANSD Working Group has elected to prepare this written briefing in order to maximize the work at tomorrow's meeting. Many of their annotations in the following notebooks will be rudimentary for the scientists on the panel, but it is important that all who see this material be able to participate as effectively as possible in the discussion.

We must not lose hope. One-third of humanity remains. It is fair to note that tomorrow's meeting is among the most important gatherings in human history. If there are questions, do not hesitate to contact your facilitator. Good luck.

BLUM'S HANDWRITTEN NOTEBOOK AND THE ANNOTATED COMMENTS FOLLOW

NOVEMBER 16, 2012

6:15 AM—ON BOARD THE AUTOMATED TRANSPORT

I'm writing this because I need to believe that there is hope.

What would be the point, otherwise? Why would I go?

I am Dr. Stanley Blum, an administrator at what's left of the Centers for Disease Control. I have a medical degree and a PhD in neuroscience, but I haven't held a test tube in over twenty years. Before the plague, I was pushing paper around my desk in Atlanta. I haven't treated a patient since medical school. I didn't do a residency. I mostly tried to stay out of the way and enjoy my family.

That was all before. If I lean over right now, I can see through the scratched window of the transport plane. The world is barren, hazy, apocalyptic. Only a fool would have hope.

But Gutierrez is the smartest person I've ever met. If we're right, if she is on to something, then there is hope. And someone has to write about it. I've seen enough computer malfunctions over the last year to know better than to type this on my laptop. I will write everything down. I don't plan to survive this trip, but this notebook just might.

And we just might find something new, something we can use, something we didn't understand or didn't even know before...Something that can cure us.

Something that can save us.

In other words, we need to start over.

Gutierrez's plan was to dissect each humanoid as if it were the first autopsy ever done by anyone. She was going to act as if we'd never seen them before. This was the only way, the only real hope that we would find anything useful, something we didn't already know, something we hadn't already tried.

And it all has to be written down.

I will record everything. Everything I see, every theory we consider, every success, every failure. There might not be any scientists left when—or if—anyone ever reads this. This might be all anyone will have to make sense of this mess. So it has to make sense to everyone. Because everyone could be anyone.

8:00 AM

STILL ON BOARD THE TRANSPORT PLANE.

Trying to organize my thoughts. Can't look at the textbooks and manuscripts anymore—my eyes hurt and I'm tired. Here's what we know so far:

- The natural history of the infection varies greatly, with time from contact to symptoms anywhere from two or three minutes to up to seven days. Classifiers at the WHO have cataloged the natural progression of disease as follows:

  o Stage I—Onset of extreme hunger with coexisting fever and upper respiratory symptoms. Cognitive lucidity is maintained. Stage I lasts from two minutes to three days.

  o Stage II—Worsening fever with measured temperatures up to 105°F. Cough worsens and cognitive decline begins with confusion and delirium. Hunger intensifies, with a preference for large, living organisms. Balance begins to suffer, with a wide, staggered gait. Arms are often held in front to maintain posture. Names and people are usually still recognized but with significant

Stage I   Stage II

Stage III   Stage IV

confusion. Stage II lasts from one to twenty-four hours.

○ Stage III—Ongoing fever, worsening delirium, and significant cognitive decline. Neurological effects are profound, with frequent falls and increased aggressive behavior. From observation, hunger appears to increase, though subjects are no longer able to speak coherently and cannot accurately convey their feelings. Previous autopsies have demonstrated significant abnormalities in the gastrointestinal organs. Malabsorption and inability to process food predominate. Stage III lasts no more than four hours.

○ Stage IV—Complete loss of human characteristics. Officially categorized as "No Longer Human"—"NLH"—by the UN and the WHO. These are the true zombies. That's what we called them when we first tried to describe their behavior, and the term, though crass, works. That's what they are… They have no capacity to recognize others as anything other than prey. They demonstrate profoundly hyperaggressive behavior. Everything that's human is gone.

- The zombie bug is a virus, related to seasonal flu, a strain of influenza. It is spherical and not segmental. This means that electron micrographs show that the virus is round, not rectangular. Spherical influenza viruses—the ones that are round in shape—are substantially more consistent with manufactured laboratory strains and seen much less commonly in nature. This is probably why Blanca thinks this bug might not be an accident. Someone may have made this thing. A bug this virulent, this tough, this scary: It wouldn't happen by accident.

- Unlike seasonal flu, this bug is unusually capable of flourishing outside the lungs. And just like the common cold, it is largely transmitted via respiratory droplets. We didn't know this for sure a year ago, but we suspected as much. Now we know. Every cough, every sneeze, every wheeze spews millions of virions. And each virion is a single, infective virus, a zombie waiting to happen.

- There are prions in the body of the virus. The word is actually a combination of two words: "protein" and "infection." They're called prions

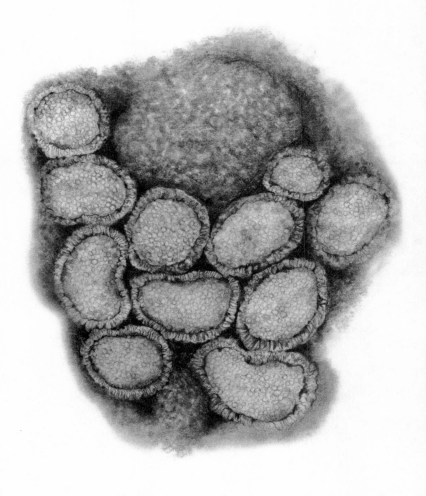

ANSD virus

because no one quite understands what they are other than infective proteins.

- But things that infect are alive, and proteins are not living organisms. This is one of the oldest mysteries of modern medicine. How can something that isn't alive infect and reproduce?

- Prions can cause spongiform changes to mammalian brains, and they're usually contracted by humans through direct oral ingestion. The changes are called spongiform because transformed brain tissue literally looks like a sponge, like mush.

- The prions in ANSD do not affect all brain regions equally—also unusual in prion behavior. For unclear reasons, the amygdala, and especially the hypothalamus, are preserved.

- We do not have a good understanding of how the prions make their way to the central nervous system. How do they get from the lungs to the brain?

- Vaccine trials have involved both attenuated (or harmless) live viruses as well as nonliving antigen-based compounds, which are components of the virus itself. With the live virus, we kept altering it, changing its molecular structure in ways that ought to have made it only mildly pathogenic but still infective enough to generate antibodies. This is routine for vaccine development. Both methods theoretically develop immunity in the host. Our trials failed, though. In fact, what we saw with the altered live virus was terrifying. The bug always reverted, always "knew" how to fix itself. It was as if it was expecting us to change it and was already programmed to change itself back. So in most cases, there was temporary success, and in all cases eventual failure. To date, no cure or vaccine exists.

- Although the molecular weights of identical influenza virions can vary enormously, there are continual reports that the measured weight of ANSD virions is greater than we can account for given the suspected mass of this particular virus and the associated prions. Is there another protein or set of proteins present? Has another virus combined with this agent?

NOVEMBER 16, 2012

2:47 PM

The situation is so much worse than I expected. I tried to prepare myself, to imagine what I'd find when I landed.

But I didn't expect this.

I stepped off the plane an hour ago. No other planes have arrived. No communications. No contact.

This is a nightmare, a bad dream.

It smells like death here.

Pittman is feverish, coughing, staggering. Gutierrez not much better. Gupta and three guards are already turned, Stage IV disease, zombies... they're gone. I counted four subjects in the holding facility. That's what's left of them.

Communication with the UN is shaky, and I can't be sure that my transmissions are getting through. I mostly get error messages and will continue writing by hand.

Still no sign of Johnson or Martinez. This was the rest of my team. I can't do this without them. I'm here to facilitate. We need their expertise.

We all left on separate transport units. We were

all supposed to be here by now. Is it possible they didn't make it?

There are lots of things out there that can bring a plane down. We knew this was a risk.

Pittman tells me that there hasn't been a food delivery for the last five days. A few droids in the sky but they didn't stop, didn't drop anything. The supplies we were bringing were loaded in Martinez's transport…We were so rushed, so desperate.

The humanoids in the holding area are hungry; we need them functional for as long as possible. They'll need to feed.

Water is scarce; desalination is continuing to function, but only intermittently. Food is mostly canned. We've tried feeding our food to the humanoids, but they don't seem interested. The stench of dead fish is pervasive. The humanoids will eat the occasional live fish that wash ashore. I guess that's how hungry they are. Usually they prefer human tissue.

Pittman and Gutierrez are infected, probably dying. They're able to work at most three to four hours a day. Pittman had two seizures yesterday. Most of their scientific notes are not intelligible, at least not to me. I've used the satellite Internet to transmit what I could back to base for assistance, but no response so far. I'm not even sure the transmissions are getting through.

It is clear that they haven't started any autopsies yet.

Gutierrez wants me to perform the autopsies. She and Pittman are too sick, too unreliable. There really aren't any other options, but the task seems almost impossible. I haven't done research in over two decades. This isn't why I'm here.

Gutierrez still suspects an additional pathogen. She suspects that our vaccines have failed because an additional bug, probably a virus, has combined with the ANSD bug. We've always been good at making vaccines for influenza, but if this bug has something more, something even more than the prions, then we're missing it and we're not vaccinating against it. If there is something more and we find it, we could figure out how to stop it. It's a good hypothesis, but the search will take time. Gutierrez and Pittman don't have that kind of time. Pretty soon I won't, either.

Until my arrival, they worried that without Gupta they lacked the manpower to manage the living dissections. Now that I'm here, they want me to lead the effort.

And then there's the message. I know Gutierrez

sent it. It sounded like she was already on to something, like she knew something new.

I asked her about it, even played it for her on the computer speakers.

Status...gent. Hype........new. Diff...virus.
Vaccine...ble

I played it three times, but she had no recollection, said she never sent it.

Is she that far gone? Is she delirious? Will she remember later?

Can I trust her judgment?

I asked her if she had new ideas, and she smiled but told me she was too tired to talk. I'm not certain that I can trust what she says. She falls asleep at her desk, often spontaneously. I have to tease out her brilliance from what looks like delirium, and Pittman isn't much help with all this.

But his drawings are invaluable. His sketches of the laboratory and of all the facilities on the island have been more helpful than any of the photographs they showed me back at the UN. I've included his sketch of the laboratory here to orient whoever reads this. So much has changed since the original construction. I also asked him to draw the stages of the disease when I arrived so we could have a working record of ANSD progression. He nodded, like he was expecting the request, and it only took him a few minutes. He's seen these changes way too often. Probably dreams about them in detail. He also drew the virus I've included, copied it from some of the electron micrographs in the island library. I want anyone reading this journal to see the enemy as it exists. It's a virus, not a person. It's a disease we're fighting. We're not fighting one another.

Pittman's drawings are the most accurate visual depictions we have.

I told him that digital records aren't trustworthy, that computer infrastructure is too shaky. I've asked him to sketch as much as he can and I'll tape his sketches into this notebook. He's a talented medical illustrator, and I need him healthy. Or at least healthier, healthy enough.

All Gutierrez could tell me is that I need to perform autopsies on the remaining subjects, and she insisted that I start with the brain. I reminded her that I am here as an administrator, but she told me that no one else on the island could reliably do this work.

She's right.

She was adamant that we start soon. She needs to see brain sections and she asked especially for

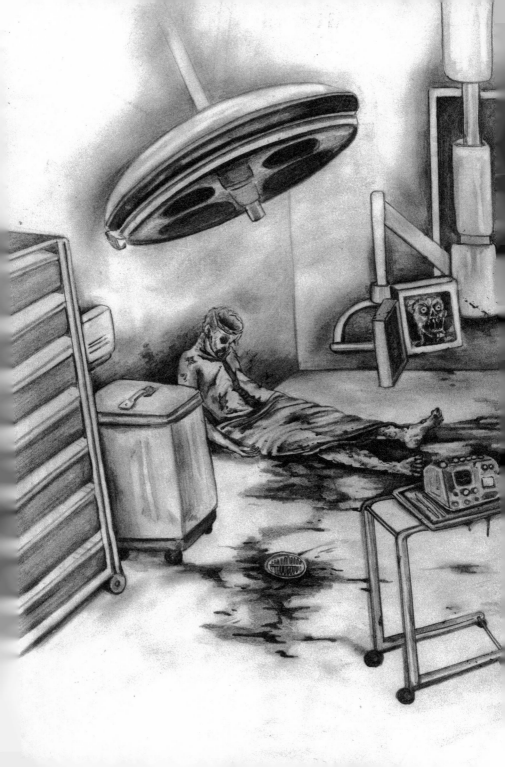

hypothalamic specimens. Because she was unsure as to her ability throughout the next few days, she outlined the following necessary steps:

- We have four humanoids (or "zombies") in the holding facility. We should start with a dissection of the brain tissue of the most recently turned specimen. We need a body that still "remembers" what it was like to be human. We're all infected, but we can't dissect one another. I remain firm in my belief that it is unethical to dissect anything other than Stage IV subjects. At the same time, we want subjects who have turned most recently so we can apply whatever we find to ourselves, to everyone who hasn't changed yet. To everyone who is still human.

- We need to study the hypothalamus, especially as it relates to the rest of the brain structures. This is a primitive region of the brain that, among other things, tells us whether we've eaten enough. Zombies never seem to have eaten enough. That's why Blanca finds this brain structure so intriguing.

- We also need to secure sections of brain material for genetic mapping and molecular characteristics. If we're going to find a vaccine, some kind of treatment, then we need to understand the exact shape of every protein, every molecule, every aspect of its genetic structure. It has to have a weak spot, a hole in its armor. Every organism has an Achilles' heel. We just need to keep looking.

- We need to understand why the heart and the lungs still function despite such severe brain disease. Most humans with brains this damaged require intensive life support. Before this plague, no one this ill ever survived. There's clearly something going on here that we don't understand.

- We need to pay special attention to the humanoids' extraordinary barriers against infection. Why don't zombies succumb to everything else infecting them? They certainly look sick. Earlier studies show an average of twenty-five or more opportunistic infectious agents throughout the body of Stage IV humanoids—each of these organisms alone would quickly debilitate even the healthiest humans, but zombies keep going. Does their skin offer protection? Are there symbiotic organisms on the skin, or maybe even the mucous membranes, that could prevent other organisms from destroying the host?

- We need to understand the gastrointestinal tract. Why don't they gain weight? Why do they keep eating? Do the nerves in the gut, the ones that talk to the hypothalamus, fail to convey the signal that they've eaten enough? Is food even absorbed? All of this will offer potential clues to different pathogens, different infectious agents yet to be considered. And a different pathogen might have different vulnerabilities, different ways to kill it. More important, if there is an additional virus, we'd need to define our vaccines with this virus in mind. Maybe the vaccines don't work because we're not vaccinating against everything that makes up the disease. Does malabsorption, the decreased capacity for nutrients to be absorbed from the gastrointestinal tract into the bloodstream, account for the intense hunger? Careful dissection and examination of the large and small bowels will potentially answer these questions. Maybe we can make them less hungry, less aggressive.

- For all organs removed, we need microscopic, genetic, and immunologic analysis. Even the tiniest observations could offer clues. Each new observation offers hope that we missed something important that we understand better now. Each new discovery brings us closer to a viable cure.

Because here's the thing: All these goals point in the same direction. They suggest that there are fundamental aspects of zombie physiology that we don't understand. Something makes them live when they ought to die. Something makes them eat when they ought to stop. Something makes them able to tolerate some of the worst infections on Earth and keep going. If this something is one thing, one pathogen, one protein, one change in the way we understand the zombie sickness, then maybe we could take that one thing away. Maybe we could treat it, vaccinate against it, get rid of it. Maybe we could cure it. Cure us.

Maybe we could save ourselves.

I plan to start reviewing autopsy procedures tonight and will examine one of the most recently turned humanoids tomorrow morning. Gutierrez and Pittman will help where they can.

NOVEMBER 17, 2012

6:46 AM

First autopsy today. Gutierrez to assist and Pittman will keep hand-sketched records. I do not trust the digital equipment for accurate video recording. Gutierrez showed me how to maneuver subjects using the transport poles.[3] We moved what was once Ben Mahoney, a former military attaché who entered Stage IV of the disease about five days ago, from the holding facility to the examination room. To date, the only means by which ANSD

---

3. The Crypt was equipped with anti-dopaminergic projectiles capable of subduing humanoids for up to ten to twelve minutes. Because dopamine is the major neurotransmitter responsible for nervous system commands for movements, dopamine-blocking agents create a state of paralysis in humanoids. That Blum chose to use the more dangerous method of moving subjects with the now-outdated transport poles suggests that he wanted as little damage to the subjects as possible before beginning the autopsy, and that he surmised that chemical damage would obscure more findings than electrical stimulation. For those unfamiliar with the early transport poles, they were titanium-based ten-foot-long hollow tubes with embedded batteries capable of producing up to one million volts through claw-like appendages attached at one end. By comparison, the average stun gun carries between 100,000 and 650,000 volts. To move humanoids, the appendage is clamped around the subject's neck and electricity is discharged. The voltage creates temporary partial paralysis, but many subjects retain movement and have been known to break free of the appendage. When temporary paralysis occurs, it lasts from three to seven minutes. As humanoids can still walk, the pole is then used to move subjects unrestrained from one area to another, though protocol calls for legs to remain shackled.

patients can be killed is through destruction of the brain. However, we need the subject's brain grossly intact for examination, especially given Blanca's insistence that we go back and examine the hypothalamus. She keeps wondering why they're always hungry, why they keep eating, what keeps that drive alive. I agree, it doesn't make sense... Anything this sick won't eat, can't eat, usually turns its back on food. Humanoids, though, drag their own decomposing bodies toward anything that moves. They never stop.

A new virus...that's all she says. It's what she said last night. A new virus. Pittman just stares when I ask him what he thinks. They're zombies, he says. They're hungry.

But on this issue Blanca is clear: We need to see the hypothalamus at work, we need the subjects animate during the dissections. In fact, I'm afraid that we all agree that the subjects must be entirely animate and conscious during our investigations. Their hearts have to be beating, their lungs breathing, their bodies functioning. This virus moves so fast that we need to observe as it works. This is really the only way we can learn enough to destroy it.

Gutierrez insisted that once Mahoney was

secure, we could start to remove the crown without any added sedatives or analgesics for pain. The Treaty of Atlanta specifically dictates that experimentation on animate zombies, without pain medications, is the best and most effective way to study the disease. Anti-pain medications might mask neurological findings. In fact, the treaty states clearly that Stage IV humanoids are no longer human, that whatever was human is gone, dead, entitled to funeral observances and mourning. We don't even afford the same rights to zombies that we do to lab rats—not because we want to hurt them, but because time is short and enacting those rights and safeguards takes time. We don't have that kind of time. We don't have time for anesthesia or pain. We don't have time for anything, really, except the task at hand. I've often wondered whether the dry language of that treaty reflects the disgust and repulsion we feel when we see what we can become so quickly, so completely.[4]

I am of course familiar with the Treaty of Atlanta and the subsequent protocol for the study of ANSD, but I still find it very difficult to bring the saw to the top of the humanoid's head. We will use a Stryker

---

4. Note that the Ecumenical Treaty of Atlanta is in appendix III of this document.

saw to remove the skull crown without damaging brain material.[5]

7:35 AM

Gutierrez seems sharp today, appears to make sense. She has propped herself up on crutches that we found in the infirmary, and she is staring over my shoulder as I try to hold my hands steady. She's still febrile—I can hear her breathing, feel how warm she is. But she looks better. Starting the work may have helped. But I need to watch her carefully. Need her to stay like this. I'm lost without her.

Gutierrez showed me how to use the spatula, prying its blade between the crumbling bone along the incision lines from the saw. Just like opening

---

5. To better understand the tools Blum chooses, one should understand that Stryker saws cut only hardened material while sparing soft tissue. Anyone who has ever had a cast removed has marveled at the ability of these tools to cut through the plaster and spare the skin beneath. However, even with the implementation of this instrument, the zombie skull crumbles. We suspect that the skull disintegrates because of ongoing decalcification—loss of calcium in bone structure characteristic of severe illness or malnutrition—among ANSD patients. This process, combined with increased intracranial pressure, explains the vulnerability of zombies to cranial injury. To date, the only sure way to deanimate a Stage IV humanoid is to destroy its brain, and the brain of a zombie is itself vulnerable to destruction precisely because the skull is decayed and at the same time almost bursting open from intracranial tension. As we learned early on in the crisis, relatively low-impact blows to the head are usually sufficient to achieve deanimation.

a can with an old-fashioned can opener, she said. It didn't feel that way, though. It felt barbaric, surreal.

The crown of the skull is now removed. Unclear if humanoid showed signs of pain. It continued to struggle against the restraints even after the brain was exposed, but struggling may be more related to restraints and hunger rather than discomfort.

The crown crumbled after it was half removed. The sutures were widened and the spaces between the sulci and gyri were large; Gutierrez explained that increased intracranial pressure was responsible for the poor skull composition.

I remember some of this from medical school. The skull comes together like a jigsaw puzzle. With a newborn infant, you can feel the soft spots, the places where the bones still need to meet. Pretty quickly, though, the bones seal. The brain has a hundred million cells and a hundred trillion connections. It is complex, intricate, and intensely vulnerable. Our skulls are fortresses, and the sutures, the lines where the bones that make up this fortress come together, are sealed and locked at an early age. It takes immense pressure to pry them apart once they've decided to close.

A healthy human brain is packed almost

impossibly tightly inside this fortress. The relatively tiny space available inside the cranial cavity is in fact maximized by the endless twists and turns of brain tissue itself. All those worm-like shapes, the bending and rebending of gray and white matter—that's how something as complex as the central nervous system can fit into something as small as a human head.

The twists are called gyri; the spaces in between are the sulci. You can barely appreciate the sulci in healthy neural tissue—they're dark and compact, each overlapping gyrus tightly pressing against the ones above and below. The human cerebral cortex is an evolutionary triumph: efficient, calculating, abstract, and awesome.

But the zombie brain is a perversion. Everything that makes us human is shriveled or lost. The dark spaces between the gyri, the once barely visible sulci, are now gaping, widened, each dark crack looking more like a grimace than a biological miracle. It's like the brain of this thing is mocking me, mocking all of us.

When we removed the crown of the skull, the top parts of the brain, in this case what was left of the frontal lobe, surged upward, as if trying to escape incarceration. I understand that this is due to the

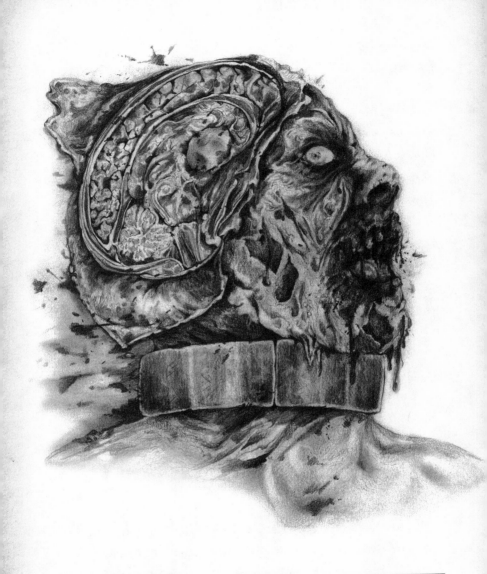

Note the widened sulci suggesting profound cortical deterioration.

same increased intracranial pressure that widened the sutures of the skull itself, but somehow I still wasn't ready for this. It was barely recognizable as neural tissue; it looked more like rotten fruit. Gutierrez just nodded. This always happens, she said. It's what we expect.

When the brain was exposed, I could literally feel its heat. Inflamed tissue is hot; that's one of the cardinal signs of inflammation, but I've never experienced anything like this. It was like opening an oven, when the heat rushes out and fills the room. We checked the monitor, and noted that the subject had increased its fever to 106°. The inflammatory response from the dissection must be exacerbated. I looked at Gutierrez and she nodded again, stone-faced, determined. This is what happens, she said. And I realized she was trying to reassure me, to let me know that so far things were going as planned. She was present at some of the first dissections, before the Crypt was in operation. She was looking for something else, it seemed, because she didn't seem surprised by anything so far.

She did, however, have her eyes on me. Describe everything, she said. We don't know what we might be missing. Don't leave anything out. She kept me on task, even as she knew, as we both knew, that her brain must look at least something like the mess on the table.

I stared at the frontal lobe, the region of the brain perhaps most responsible for our humanity. This is where we reason through problems; make sense out of base emotions, decide whether we want to behave in a way that we'll later be proud of or perhaps come to regret.

But not here, not what I was staring at. The tissue almost immediately lost form, seemed partially liquefied as it seeped over the sides of the open cranial cavity. Forgive me—it looked like old meat left out too long. Something that might attract flies or a homeless cat. Something that in simpler times we would find on our driveways and sidewalks, that we would pick up with a plastic bag, careful to breathe through our mouths or perhaps holding our breath altogether, and toss without a thought into the sealed darkness of the garbage container... Something foul and insignificant.

These are the spongiform changes brought about by the prion component of the disease. I've seen photographs, drawings. In a sense, I knew what to expect, but photographs don't tell the whole story. This stuff moves. The brain material literally changes shape as we expose it. It pulses with each beat of the subject's heart, and each pulse forces more tissue up and over the walls of the opened skull.

As I described before, the sulci are widened, and the vessels that bring blood to the brain, the neuronal vasculature, are engorged from the increased pressure. Inflammation is one of the most fundamental responses to tissue damage. It conjures up fever and blood cells, defenses against something that doesn't belong. The fever burns, the heart beats faster, and the vessels engorge with the increased pressure.

That's why the brain of a zombie isn't all gray. The vessels can't hold this kind of pressure, and even as we watched, vessels were actively bursting. Bright red blood mixed with the brain as it dripped onto the gurney and into the gutters at the sides of the table. Blanca again had to focus me, remind me to obtain samples of material before they were contaminated by touching the table or the floor. We tried, but the best we could get was a soggy mass of something no one would ever recognize as anything close to brain tissue. And covering the tissue was a gelatinous substance, oddly translucent and seeming to glow from the glare of the operating lamps. This is the end result of brain degradation—decomposed tissue, broken down by bacteria that have attacked the brain as it becomes increasingly vulnerable.

Throughout all of this, Blanca looked just beyond the surgical field. It was as if she was thinking of something. I realized again that she has seen all this before, that she was still looking for something new.

It's hard not to think of this thing as alive. The brain itself continues to pulse. As we removed brain tissue, we could visualize the optic nerve. In healthy humans, the eyes are literally holes, giving us a glimpse into the darkness of our skulls. That's what I used to say in classroom lectures before I left academics and went to the CDC. I can see your brain right now, I'd mention, and then I'd project a slide of beautiful eyes on the screen at the front of the classroom. Our pupils are dark because they are connected directly to neuronal material. The optic nerves look like elongated tissue with eyeballs attached at one end and the rest of the brain at the other.

And here's what was strange. The optic nerves seemed fine, perhaps even bigger than expected. Whatever damage the prions did to the frontal lobes was sparing the optic nerves, which makes sense, I guess, because the damn things can still see. I made a mental note to examine the occipital lobes as well.

This is the region of the brain where visual input is stored. I had read that the prions involved in ANSD also seem to spare occipital activity, again explaining the ongoing function of the visual apparatus.

The muscles that control the mouth were also more or less intact. There was some damage, some signs of infection and degradation, but zombies need to feed, and even as we separated the brain from the body, the subject kept snapping its mouth open and shut, trying to bite something, anything, the muscles responsible for these movements contracting and retracting with machine-like precision.

Finally, there's the stench. It's rotten, like something already dead, but mixed with the smell of iron and copper. This is the blood and the tissue, intermixing and decomposing, actively disintegrating as the disease progresses. It's why some have reported smelling Stage IV humanoids even before seeing them. The smell is rancid and, over the last couple of years, horribly familiar.

So far this is all consistent with what we've seen in past studies. Although we know that scientists working at the Crypt may have genetically manipulated the virus in order to better examine its different properties, there is no sign that Mahoney's infection is any different from what has been

Note the activity of muscles of mastication that seem unaffected by ongoing dissection.

previously recorded in earlier examinations. This is good. We need to know that what we're studying is consistent with the disease as it exists off this island. From what I've read, and Gutierrez agrees: This is the brain of a zombie. This is "normal."

As expected, the quality of brain tissue is sufficiently poor that we'll need to use the spatula to get deep enough to examine the hypothalamus.

8:50 AM

Gutierrez just passed out. Pittman helped me to get her back to the bunker. We suspect dehydration. Subject is still in restraints with the crown removed and the brain exposed. It should continue to be animate—most of the initial neuronal seepage seems over, and the loose tissue has either been removed for samples or remains in the gutters of the table. As long as we can feed it, it should be OK. But right now, we need to nurse Gutierrez. She's still unconscious. I'll return to the lab later tonight or tomorrow once I'm comfortable Gutierrez is stable.

So far, our findings reveal nothing new. We expected gross spongiform changes, the prion-induced degradation of brain tissue, and we expected these changes to be most profound in the frontal lobe, in the higher brain structures. This is in fact the pathologic basis for the spiritual conclusions of the Treaty of Atlanta. The frontal lobe makes us human, and its destruction robs us of our humanity.

Still, it's hard not to mention that the degree of brain tissue disintegration is alarming, even horrifying, to directly observe. We need to be careful as we proceed. Stay focused and detached. Brain tissue is so fragile that we run the risk of contaminating any hypothalamic specimens during removal of gray matter.

It is also not surprising that the subject retained consciousness, though again it is unsettling to observe this phenomenon. All it needs is the brain stem, the part of the brain responsible for automatic functions. As long as the brain stem remains functional, tells the heart to beat and the lungs to breathe, zombies stay animate. They don't need much more.

After Gutierrez is recovered, we will resume excavation of the brain with the primary goal of isolating hypothalamic material. I need to think about this. When I mentioned the hypothalamus to Gutierrez last night, she became excited, even agitated. This is where the answer is, she said. It's what we missed. They eat and are never sated. Why?

The hypothalamus should tell them when to stop, should remind them that they are ill, should rob them of their appetite, and yet for some reason their appetites are endless, reptilian. There are infections that can do this, she explained. Some viruses invade the satiety sensors of the hypothalamus, confuse those sensors, and interfere with the way they talk to the rest of the brain. Some infections can make people think that their stomachs are empty no matter how much they eat. These people are always hungry…always wanting. People with these viruses are relentless.

What if one of these viruses is part of all this? We've never looked for this kind of hypothalamic infection before. We haven't even looked. We don't have time to miss anything. How can we afford not to look?

From this point forward, we'll proceed according to strict protocol. We'll finish this dissection of the brain and then start to move downward, studying both microscopically and grossly the cardiac, pulmonary, and gastrointestinal systems. If a subject becomes inanimate during the autopsy, we'll move to the next one in the holding facility and pick up where we left off.

We'll have three main goals:

1. We need to find evidence for Blanca's theory of an additional organism. This is our primary goal, our main hope.

2. We need to better understand the barriers against infection that may exist among humanoids and make those who are NLH particularly resistant to biological attack. This might give us some additional clues about how best to attack the virus itself.

3. We need to stay alert, vigilant, for anything that may have been missed in previous studies.

Only by sticking to protocol, by systematic and careful examination, can we do this right. The clock is ticking.

NOVEMBER 18, 2012

3:34 PM

We desperately need to continue, to finish the brain dissection and study the entire organism. But Gutierrez is still sleeping, getting sicker. I can't do this without her. And the subject—it can't remain animate forever. It's tied to a steel stretcher, restrained on a gurney in the lab.

It's like us now. Trapped and hungry. It needs to feed.

Time is so short—I've tried to imagine continuing without Blanca and her guidance. What if I miss something that she'd notice? She's a good scientist. She at least has some idea of what she's looking for, but I've read every document on her computer. I've studied her sketches, looked through everything she's written. I've even pored over the letters she used to write to the long-dead pastor in the village where she was born. She's seems to be trying to organize her thinking, but her thinking is what worries me.

She keeps coming back to a "novel virus," and the idea that the ANSD pathogen is something new, something we haven't seen. It's more than a combination of influenza and prions. After all, we understand influenza, and we pretty much understand prions as well. We might not be able to stop prions, but if the prions are delivered via the influenza virus—if the vector for the prions is itself an influenza bug—then it stands to reason that conventional anti-flu treatments should have at least some effectiveness. And yet, other than slowing the rapidity of prion damage by altering our internal pH, we've been powerless.

Maybe a new toxin, some kind of poison secreted by the ANSD bug that we haven't yet discovered? She has all these drawings of molecular structures that I don't recognize, can't even begin to place. Maybe she's wondering what kind of toxin, what kinds of proteins a novel virus would produce, what it could secrete that we'd be so unable to stop.

But then she'll shift gears, right in the middle of the page, and suddenly she's back to old stuff, back to manipulating the pH, back to trying tired anti-influenza medications with slight modifications. She's all over the place, and none of it makes any real sense. At least not to me.

She's confused. Or she's on to something very new.

Probably both. Either way, I need her.

I need to wait until she's recovered enough to assist with further exploration. We don't have unlimited subjects, and I can't do genetic inquiries and molecular studies without a better guiding theory. Dammit, I can't even remember how to properly conduct a protein analysis. I'd be more comfortable if we had a thousand needles for this haystack, but we don't. We have four. Four subjects and a bunch of garbled theories that no one except maybe Blanca understands, which means that for now she's our best hope.

But to maintain the integrity of our investigation, the humanoid in the laboratory still needs to feed. Its metabolic rate is so high, it's starving. To preserve nutritional status, at approximately 11:35 PM I fed it the recently removed left ear of another subject from the holding area.

The wind was blowing, and the smell was awful. The floodlights that illuminate the holding facility were swaying at the tops of their steel poles like fruit on old trees, so the light kept shifting, casting shadows through the mist. It was so dark, so strange. I was entering a cage, felt trapped.

One of the subjects, a female I think, was on the ground, resting maybe, her face staring upward toward the sky. As soon as I took my first step into the facility she moved, was on her feet and lunging. The chains held, though. Thank God.

I used one of the transport poles to immobilize her and used the emergency restraints in the holding area to secure her head. Her skin was worse than the one in the lab, the tissue on her face seeming to dissolve with the forces of gravity.

But it knew I was here. It was decaying, actively

decomposing even as I watched it, and yet it struggled with ferocious effort. It seemed not to care at all that I was removing its ear. It only wanted to be free.

How much energy can one creature glean from the limited nutritional value of badly damaged cartilage and a few disjointed proteins? How many more times will it need to feed? And what on Earth are we going to feed it?

I am trying to be as objective as possible, to dispassionately include every detail. Still, I'm offering "food" to something I've been told is no longer alive, or at least NLH.

And removing the ear from the subject in the holding area was even worse; it showed no emotion other than rage, and twice it bit my Kevlar gloves. How do we do this? How do we maintain our sanity? I just fed an ear to a zombie.

Gutierrez woke slowly, asked for water. She vomited after a few sips and went back to sleep. Soon she was awake again, this time even able to hold down some food. She woke hungry.

I'm OK, she told us. We have work to do, and she instructed me to check on the subject. She was shuffling toward the closet, getting the wheelchair, unfolding it, and lowering herself into the seat.

All three of us went back to the lab and found

Removal of left ear from subject in holding area.

that the autopsy subject was resting until we entered the room. In fact, even before we arrived, it appeared to sense our presence and started vocalizing.

The other subjects in the holding area heard the vocalizations and were responding. The sound of their calls filled the air, but were muffled by the wind, the white noise of the dead ocean.

Once we were in the laboratory, the subject became quiet for a moment—it strained to see us, its pupils tracking our movements even as its brain remained exposed. Its head was strapped to the table but it could move its eyes. It was watching us.

Then it started growling.

Much of the frontal lobe had entirely liquefied, suggesting little use for this region of the brain once humanoids are totally infected. Blood vessels had burst throughout the night, and there was blood on the wall near the subject and pooling on the plastic floor.

For God's sake, this is the frontal lobe. This is the higher brain, the human brain, the marvel of the known universe. This is the part of our brain that makes us wonder, makes us dream, makes us pray.

Couldn't the plague have spared this? It could have taken our movements, wrecked our senses, blinded us and made us mute, but, please, let us keep what makes us human. All that stuff about divine retribution. All those people preaching that this virus was the end, something sent from above, a second deluge, a punishment for what we've become.

We were all thinking that, and what's left of humanity still does, for the most part. But now... now, looking at this, I know this has to be wrong. This plague was created by something more primitive, more base. Only a person, a terrible, misguided fellow person, could stoop low enough to create a virus this darkly perfect. This virus is rank, made by something evil, something human.

Gutierrez just asked me why I was writing so much—she reminded me to focus, to be dispassionate. I'm sweating, though the climate control is functional right now. I'm overwhelmed.

No. That's not it.

I'm angry.

I am looking at the frontal lobe and it is indistinguishable from decomposed matter; it actually smells dead. I can't even grasp and maneuver it. The tissue that remains intact is anchored by the brain stem as it attaches to the spinal column, but whatever I grab with my hands, even with the forceps, almost immediately disintegrates. It's like melting snow. It drips through the cracks in between my fingers.

Pittman is sketching, Gutierrez is taking notes, and I need to focus. This is what a neuroanatomist would see. This is what I can observe:

- The sulci are so wide that there is an absence of gyrations altogether. The fundamental architecture of the frontal lobe is gone, missing. It's like soup.

- The occipital lobe is also affected, but, interestingly, it is substantially more defined than the rest of the visible brain structures. I always thought it strange in normal human anatomy that the occipital lobe should be so far from the eyes themselves. This is the region of the brain that processes visual input. A person with a badly damaged occipital lobe might as well not have eyes at all. That's why the back of the skull is so solid. How many times do we fall down on the ice, slip on the sand, slam the back of our head onto cold concrete? Damage the occipital lobe and we risk sight itself, so our skulls are extra-thick, extra-careful to preserve this fundamental sensory experience.

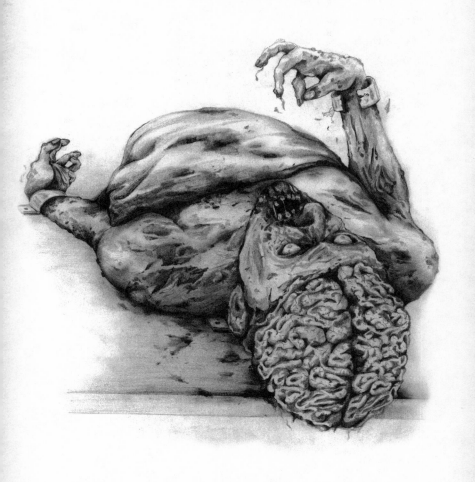

Note the widened sulci and loss of neuronal architecture.

And there is no doubt about zombies—they can see me, can see all of us. This one even tracks my movements as I walk back to my stool to write.

We're moving quickly, but I need to concentrate. I don't want to miss anything.

I had to remove the higher brain with a spatula. It was quickest this way, and we need to dig down, get to the hypothalamus as quickly as possible. Surgeons often approach the hypothalamus through the mouth, cutting through the top of the palate, taking a shortcut to the deeper brain structures. We considered this, but our subject must be animate and awake. We need to watch the virus at work, as it actively infects and changes brain function and structure. Anytime these things are animate, they're biting, chewing, acting as if food were there even when the mouth is empty. We'd lose our way quickly with that kind of movement.

So we dug down from the top, removing the brain the way you remove soil under a tree to examine the roots. I could sense Blanca's excitement as the thalamic structures came into view. Below them, she said. It's called the hypothalamus for a reason. "Hypo-" means "below." The hypothalamus sits below the thalamus. Easy to lose your orientation when the brain itself is so damaged, so these words have special meaning now. I carefully retracted more tissue, dug just a little bit deeper, and Blanca smiled.

Some of the hypothalamus seemed preserved, appeared grossly normal and even larger than would be expected. It was as if there were a fence around it, stopping the spongiform changes the way an electric fence holds back an angry dog. To slow something as virulent as the ANSD prions, there would have to be something even more powerful. It would have to alter the pH of the surrounding tissue maybe, make the surroundings more basic, more alkalotic. Something very significant was protecting at least some of the hypothalamus from this plague... We haven't figured out a good way to stop this disease from spreading throughout the brain, but whatever was in the hypothalamus can. In fact, it seems to me that only another microorganism could do this. Only a separate contagion could even begin this fight, and that means the hypothalamus is infected by something different that we haven't even considered yet. This is what Blanca was talking about.

It was tricky to extract the hypothalamus. I had to secure it with forceps and peel the less-formed tissue out of the way. No way to do this except with my fingers and a few instruments. Finally it was

Hypothalamus quickly loses form once removed from connecting brain tissue.

loose enough, and I carefully lifted it free with the spatula for Pittman to sketch. It's hard to recognize it once out of the brain—it looks naked and ill formed. The amygdalae are still down there, buried beneath this mess, but the hypothalamus is extracted and free.

I'm trying to get my mind around this. Three contagions operating through a single vector, a single means by which the contagion can infect a given individual. Every germ has its vector—food poisoning comes from contaminated food. Skin infections use direct contact as the primary vector. Hell, until now, prions could only get inside you if you ate them, but now they're in the air. For ANSD, it's the air itself; that's our vector.

Each cough, each sneeze, each gasp for breath from an infected lung spreads an influenza virus with ANSD prions packaged neatly and efficiently inside the virus itself, but each cough might additionally spread something more. We've long known that ANSD was a combination of more than one infective agent. We knew about influenza first and later the prions.

But could there be a third infectious agent in the same "package"? Could there be something that alters the milieu of specific brain tissues? Could the hypothalamus be protected by a competing contagion? Worse, could all three contagions be symbiotic? Could they be "helping" one another?

This symbiosis would have to be designed, would have to be planned. This would be engineered carefully, even brilliantly, in a highly technical lab. Nothing like this has ever evolved on its own.

Gutierrez insisted that we immediately dissect the hypothalamus. She mentioned that some viruses attack only this brain region, but she seemed not to remember that she had already informed me of this hypothesis. She also noted that this brain structure was clearly demarcated, easy to discern. Its architecture was intact. There's something here, she said. There has to be.

Additionally, there is something quite large near the inferior, or lower, aspects of the brain that is well demarcated and seems to be pushing the thalamus and hypothalamus upward. Those intact regions maintained their integrity but they were clearly being moved, forced upward in a rhythmic way by whatever sits beneath them.

Get the sections, Gutierrez said. Get the samples, and I tried to hold my hand steady as I moved the scalpel through the hypothalamus. We need thin slices for examination. We only have one chance once the blade hits the tissue. Gutierrez tried to calm me. You've done this, Stanley, and I was thinking the same thing. I've taken plenty of brain sections, made thousands of slides.

But I haven't done this for a very long time.

I remembered what my anatomy professor taught me, now more than thirty years ago. Anatomy doesn't change. Cutting the brain is like cutting bread, he said. Be gentle and confident. Be true with the knife.

I removed four samples of the hypothalamus and passed them to Gutierrez. Two of the samples looked pretty good, the architecture intact, the tissue robust. But the other two samples, the ones from the back of the hypothalamus, looked somewhat worse. It wasn't noticeable at first, but when I made the cuts the tissue sagged, crumpled, like a tired wall. I mentioned this to Gutierrez and she looked pleased. She wheeled out of the room to see if we had enough power in the generator to work the electron microscope. A solar-powered generator can only do so much when the sky never clears, she said, and she turned to leave, stopping at the door. We might

need to power down other resources, she mumbled.
It was the first time I had seen her look concerned.
Then she turned to me. Find out what's under the
hypothalamus, she instructed. Something's down
there that's making this tissue move.

I used my fingers to explore, gently pulling at
tissue to insert my hand underneath what remained
of the brain. There was something there, without
question. It was hard, like clay. Was it a tumor? A
fungal infection?

It moves with the rhythm of the heart, rising and
falling with each cardiac contraction. Whatever
this is has got to be at least functional tissue, part of
the overall organism. And it feels as healthy as the
hypothalamus, maybe even more so.

Dammit, this is "healthy" tissue below the
hypothalamus and I can't correlate it with any
expected brain structures. It has to be highly
vascularized, it has to have ample blood and oxygen,
to be this healthy, to be this tied to the beating heart.
With a brain this sick, when even the gross pathology
is completely abnormal, anything that remains
healthy, anything that actually recruits its own
blood supply must be there for a reason. Whatever
this tissue is, it is vital to the success of the infection.

It might even be central to the behavior of the zombie.

As Pittman moved closer to study my dissection, whatever it was that I was feeling quickened its pace, seemed alive and alert. The subject flared its nostrils. Jesus, it looked like it could smell him.

Readers may be alarmed that the subject remains awake during the procedures. It is important that this feature of the process not be confused with a property unique to ANSD Stage IV humanoids. In fact, even noninfected humans in sterile environments will remain alert during open-brain procedures, and the hypersterile environment of the Crypt laboratories was presumably still intermittently functional.

Blum's findings, however, remain some of the most perplexing aspects of ANSD pathophysiology. Prion infiltration usually invades much of the brain, and in all cases of successfully examined ANSD subjects, specific brain structures remain preserved or at least relatively preserved.

If indeed a third contagion exists, its role might be to protect the brain from the rapid degradation that ANSD prions would otherwise generate. This is the first new theory in more than a year, and we're already looking at ways to investigate its validity. Maybe the key to a cure is to interrupt the protective properties of the third contagion. As Blum correctly notes, this kind of symbiosis—three contagions keeping the host alive, keeping it active, allowing it to spread the disease efficiently and widely—would almost certainly be an engineered plague. In fact, very few countries before the outbreak possessed the necessary technological knowledge to even conceptualize this disease. Our forensic teams are scanning biowarfare intelligence documents for signs of anything resembling even theoretical initiatives regarding ANSD. So far, we have found nothing to implicate any country, organization, or individual, but the process is ongoing.

5:25 PM

And that thing, that hardened mass below the hypothalamus?

I was wrong.

Not a tumor. Not infection. Robust, almost supernaturally healthy tissue.

Grotesquely healthy.

The masses are the amygdalae themselves. The crocodile brains.

We're dissecting crocodiles...crocodiles that used to be human.

We're dissecting monsters.

We had planned to keep the subject animate for the entire dissection, but Blanca feels strongly that we need brain stem tissue for analysis. Severing the brain stem will deanimate the subject, but Gutierrez reminded us that additional subjects remain for further animate dissections. We need to know now, she said, why the brain stem stays functional.

Using nine-inch serrated Yasargil neurosurgical scissors, the brain was removed from the spinal column at the C1 cervical vertebra. As expected, separation of brain stem from spine deanimated the subject, though the pupils remained pinpoint longer than expected. I suspect this was due to a heightened fight-or-flight posture even after

Removal of amygdaloid material.

initial deanimation. Adrenaline still runs high. The lack of immediate dilatation of the pupils that would be observed in normal human death is consistent with the increased sympathetic nervous system activity that characterizes Stage IV humanoids. Zombies are biologically primed for aggression at both anatomical and molecular levels. As I said, their adrenaline runs very high, even after the brains stop working.

The brain stem itself is remarkable precisely because it looks so normal. It appears entirely intact. Sections were obtained from the pons and the medulla, key components of the brain stem itself, in order to study the reticular system of the subject. They're called this because of the meshwork, the reticulum of primitive neuronal fibers that intermingle and communicate. These are the parts of the brain stem responsible for keeping us alert and aware.

I made some rudimentary slides and studied them under the microscope. They're damaged, but it's unsettling how normal these samples looked compared with the soup we had to cut through to get these structures.

There's no doubt that these things are aware. Until I separated the brain from the rest of the body, it knew we were here. It is awake the way a crocodile waits for its prey. It would have bitten me even with most of its brain removed.

Just above the brain stem, it looks like the cerebellar tissue is somewhat affected by spongiform changes but not entirely liquefied. Because the cerebellum helps with balance and coordination, this also makes sense. Anyone who's ever been drunk knows what it's like to try to walk without a healthy cerebellum. It's a remarkable structure, constantly making minuscule changes, a computer really, devoted among other things to keeping our gait fluid and smooth. One of my professors in medical school used to call the act of walking "a series of catastrophes narrowly averted." Your cerebellum averts the catastrophe, he explained. It keeps you on your feet, helps you to recover even before you're about to stumble.

ANSD tries to get at the cerebellum; I can see the damaged tissue, but the changes aren't nearly as stark as in the higher brain regions. Slower degenerative processes in the cerebellum explain the initially intact gait of the infected, even though they all become increasingly unbalanced with time. That's why they hold their arms out in front of their bodies: for balance and increased coordination. They just want to remain upright, on their feet. But the process continues, the cerebellum degrades, liquefies. Virtually all late-stage ANSD humanoids ambulate via crawling.

The nearby hippocampal tissue also looks pretty good. Sure, it doesn't look entirely healthy, but it looks functional. Because the hippocampus stores memory, this seems to explain those rare instances in which Stage IV humanoids seem to "remember." There are numerous controversial reports detailing how zombies purportedly return to where they've previously fed.

Gutierrez immediately began preparing the hypothalamus for study. She was careful to separate different hypothalamic regions—the ventral part was placed at one end of her work space, the medial at another, and the dorsal at the far end. She was sweating, working quickly. Her brow was furrowed and she kept stopping to wipe sweat from her eyes with the back of her hand.

Zombies are always hungry, she said. They're never sated. Why?

I hope to hell we find out who did this, and I hope to hell that person is still alive.

Whoever did this is as deranged as he is brilliant, and if we're lucky, we'll find him and understand the disease, his disease, that much better. Whoever made ANSD might have made an antidote as well.

Or maybe not...

We never developed a good antidote for radiation, but we made enough bombs to destroy the world about a hundred times over. Let's hope our evil scientist had something besides destruction in mind.

We dissected the eyes and optic nerves from the rest of the brain.

I placed the eyes next to a measuring tape for Pittman to sketch. They appear normal in length and thickness, so it looks like the pathogen spares the eyes, just as it does other sensory organs. Again, this tells me that whatever made this virus meant for its host to see, to track its prey, to kill. It's almost as if this were a botched attempt at creating a weapon. An army of these things could overwhelm almost anything.

Signals from the brain to the muscles of the eye are probably at least somewhat disrupted—hard to imagine what they actually see—but this at least explains why zombies often misjudge the speed of their food or the distance they need to lunge. Still, they don't need to be accurate. Just relentless. One zombie isn't a problem, but a hundred zombies are bound to catch up with you. I think we are beginning

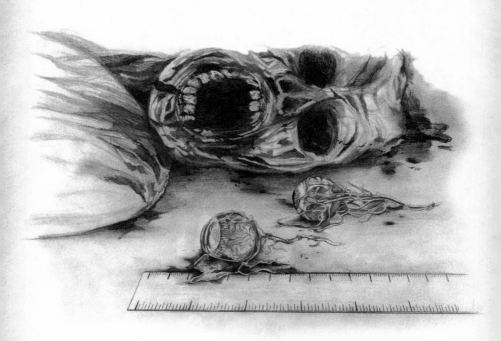

Removal of intact eyes and optic nerves.

to understand that ANSD is efficient and utilitarian. It preserves just what it needs to survive.

When we first realized the scope of the pandemic, we weren't ready. ANSD brain material was already so damaged, so frail. Conventional chemicals, routine laboratory preservatives intended to keep tissue samples intact proved too caustic for specimens taken from the infected.

We thought we could go forward right away with all the tools, with the technological knowledge that we already had. We didn't stop to consider that this was something novel, something we'd never seen before. When it became clear in those early days that our technology wasn't fit to study the outbreak, a cottage industry of new laboratory equipment and chemicals developed directly from the need to better study these things. Industries finally came up with fixatives that could preserve ANSD-infected tissue without damaging it. For a very brief time, some companies got very, very wealthy.

None of that matters now.

Tomorrow we'll open up the thorax and abdomen. Nothing says this virus only takes our brains—if we can find signs of infection, or protection for that matter, in the heart, the lungs, the gastrointestinal tract...

If the brain is this damaged, then we need to understand why the rest of the body doesn't give up. Complex organisms this sick shouldn't last. Until recently, anything whose brain looked like what we've seen would have died quickly and completely even with the most advanced life support. But these things walk! They ambulate and feed, and presumably their hearts somehow get blood to the vital organs, the lungs move air, their digestive tracts absorb at least some nutritional content. It would be good to know why—we might find areas of vulnerability that we haven't yet considered...

Think about it. Influenza usually kills by pneumonia. It usually drowns the lungs with pus and we suffocate, coughing until there's no more air to cough.

These things breathe.

They cough and they sputter, but they continue to respire. Air goes in and air goes out. How the hell does that happen with anything this sick?

I need to see the lungs. The disease is spread through respiratory droplets. We learned this early in the outbreak. People spread influenza through coughing, through blowing their nose, and in the beginning this looked just like influenza.

Except almost every strain of influenza

encounters natural immunity, and no one is immune to ANSD. It may look like influenza, but it behaves differently. Again, it behaves like someone accounted for the inherent immunities of naturally occurring viruses. It behaves like someone wanted it to be this bad.

That's why the subjects in the holding area are still animate. That's why they still want to feed.

We can use removed brain material from this one for nutrition for the others; they tend to prefer to eat central nervous system material. We may have to heat the tissue—that seems to increase the likelihood that they'll bite—but we'll do whatever it takes.

Most important, we need to keep the remaining ones "healthy."

There is a symbiotic process here, a process that allows the contagion and the host to coexist and, at least for a while, to thrive. We haven't thought about the infection like this. We've focused on how it kills us, not how it keeps us alive...

NOVEMBER 20, 2012

4:35 AM

Slept for twenty-six hours, off and on. Air-conditioning is intermittent. I am coughing, diaphoretic. My cot is soaked.

Gutierrez is in the infirmary—Pittman carried her there yesterday while I slept. She is better after IV fluids, but she is clearly dying.

Before I knew she was in the infirmary, I went to find her this morning in her room. Her quarters were empty and water had condensed onto the keyboard of her computer. The screen showed images of goats and horses and a molecular image. I'm going to ask Pittman to sketch the molecule—I think it's a virus. I can't trust her explanations—she just doesn't look right. But I'd like to study it later, when I have some time to think.

I feel fuzzy, like there's something slowing my thinking. It's horrifying, this fuzziness, because right now is when I have to be sharp, to think more clearly than ever before. I'm scared. I think this is what it feels like to go mad: to not know whether you can trust your own thoughts.

Writing helps—it is organizing, helps me to remember myself. The journal feels like oxygen.

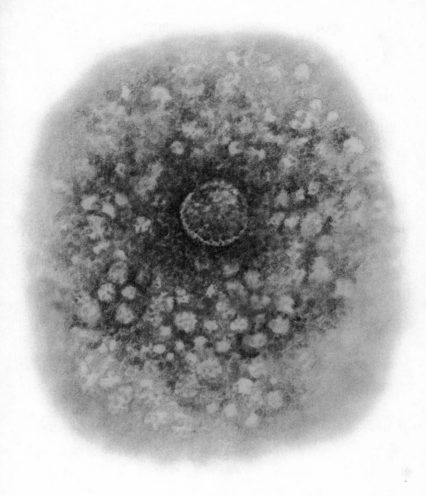

Unknown molecular structure found on Gutierrez's computer monitor.

Things seem more clear when I read what I've written.

So, here's what we've learned:

Most important, someone made this thing. The more I see through the autopsies, the more I know that this virus was built, created, engineered. Whoever reads this journal needs to find this person, the creator, and hope to hell that he or she is still alive. This wasn't made in somebody's basement. There must be records, journals, failed experiments, and maybe even lab animals that we can find and study. This has got to be a priority for whoever reads what I'm writing.[6]

Because it now seems clear to me that the pathogen does something even more grotesque. It triggers some kind of novel immune responses and refuses to let us die mercifully once we're infected and changed. It seems more and more likely that either the ANSD pathogen itself or, maybe even more likely, an additional infectious agent protects regions of the brain necessary for the behavioral changes induced by ANSD infection.

---

6. A multinational team of military and scientific experts has embarked on a worldwide search for evidence of facilities capable of creating the ANSD pathogen. We are presently trying to access functional surveillance satellites in order to better narrow our search parameters. We will continue to apprise you of these efforts.

Currently, the immune response of the influenza and prion components of the ANSD contagion do not show any signs of recruiting the necessary immunological response to prevent what ought to be much more pervasive destruction. If an additional infectious organism creates this immunoprotection, then it most likely targets the very regions of the brain that are not affected or less affected by prion infiltration.

But there's so little time. The remaining zombies could deanimate. Blanca could die. I could fail...

We need to hurry.

Unfortunately, when I went to the lab this morning, I found that the current subject is already badly decomposed. Apparently, the intermittent power outages affected air-conditioning there as well. Pittman and I turned the vacuum hoods to high pressure to clear the room of flies and other insects. There were crabs feeding on the deanimated body. We caught them with nets.

There are things alive that feed on the dead. They don't seem rushed or eager—they're just content to have found something to eat. The crabs walk sideways, their eyes waving around on twig-like stalks, their optic nerves clearly exposed. This is normal crab behavior, normal crab anatomy. Their eyes exist only to find prey. They're not rushed. They seem to know that there will always be something to feed on.

Crabs, zombies, us. Is there a difference?

And the world seemed to be spinning then. The dissections, the flies, the feeding crabs—all of it was just too much, and I was sick, angry, rattled. I vomited, and when I wiped my mouth with my shirt I saw Pittman staring. His eyes were wide, like he was seeing it all for the first time. What to do? Do I smile to reassure him? Do I ask him to sit down? How can I protect him from all this? I can't even protect myself. I can't afford even to vomit.

5:50 AM

My pH is in the safe range and I am holding off on further diuretics. I've rested, feel better. Need to get back to work. Vomiting can disrupt the alkalosis, and I can't afford to lose more water. I'm ready now to move forward. Too much time has already passed.

We need a new subject.

Three humanoids remain in the holding facility. Transport poles feel heavier than even a few days ago, but Pittman and I moved a zombie to the lab with little difficulty. Dog tags identified it as Gupta.

Christ, I know Gupta. Knew her. She is weaker than the last subject, barely recognizable from skin infections and ulcerations. I knew she was down here, knew she had turned, but this is different. I just shocked her with a transport pole.

I knew her family. We met when this all started, at the first international gathering.

6:30 AM

Humanoid broke free of the transport pole. The chains on its legs rusted and snapped. We subdued it with darts, but only after it bit Pittman's chest, lunging at him with incredible speed—maybe even skill. The relationship to paradoxical Parkinsonian movement is remarkable. People whose movements seem rigid, slowed, even paralyzed by Parkinson's disease can sometimes move with amazing fluidity when they're frightened or excited. They can't do it on command—it's all instinct, all drive. Zombies do the same thing, and we should have been ready. Even something as weak as Gupta can strike quickly, without warning.

We stopped Pittman's bleeding; it got him just under the ribs, but didn't pierce the skin much beyond the dermis. Its teeth came out in his skin.

I administered broad-spectrum antibiotics, but Pittman looks bad, shaky. I tranquilized him and left him in the infirmary resting next to Gutierrez.

Video surveillance of the lab in the infirmary shows subject still unconscious on the floor of the lab. More crabs.

I need to sit down, just for a minute, just to catch my breath.

6:40 AM

Arrived back at the lab and found the subject still unconscious but seizing. Crabs were scuttling over the body, trying to grab hold of her as she convulsed and writhed. Seizures will disrupt the integrity of the vital organs, might even obscure immune markers. Grand mal seizures are rare in ANSD until final stages; I needed therefore to stop this seizure in order to preserve the utility of the autopsy finding. I treated the humanoid intramuscularly with large-dose injections of lorazepam. This was the only anti-seizure medication I could find. The lab was a wreck from the fight earlier when the subject broke free. I know that lorazepam is a sedative, that it might slow some of the processes we need to study, but I don't think it will affect the lungs or the surrounding anatomy. And I'm not ready to sacrifice another subject.

The seizure abated in about five minutes, and the humanoid was sleeping, sedated. I hoisted it onto my back and gently laid it down onto the dissection table. I was amazed at how light she is, it is. It's barely even here. I ignored the crabs that had skittered into the corners of the lab and focused on what I had to do next.

I fastened the wrist, ankle, neck, and midline restraints. I checked to make sure they were solid. The capacity for even very impaired humanoids to attack is clear. Its breathing is steady now. Temperature is 104.3°F.

I am grateful it—she—is sedated.

Using a standard body block, I elevated the trunk and allowed the arms to fall to the side. I did not deanimate it—the head and brain are intact. We need the subject animate in order to observe the organs in vivo, while they are still "alive."

Using a long-handled autopsy scalpel, I made a Y-shaped incision, starting at each shoulder and moving down to the skin just above the sternum. Two humanoid ribs broke when I lost my balance and fell, blocking my fall with my elbows on its chest. Like the skull and scalp of the last subject, the bones are friable and fragile.

The layers of skin are decayed, the muscles rotting, the bones barely solid. But the heart is beating, its valves opening and closing just as mine are right now. As we hypothesized, ANSD leaves functional exactly what it needs.

I removed the dermal layers, placing them over the eyes of the unconscious humanoid. I don't know why I did this. When I dissected the head two days ago, I couldn't see it watching me, but if this one wakes up it'll see what I'm doing. It will look into its own thorax, and it won't recognize a thing. I think that's what rattles me the most—not that it will see me, but that it won't give a damn about what it's seeing. It will stare at its own beating heart with dead eyes, and I don't think I could watch that incongruity. I think seeing that might push me over the edge.

The fact that the heart is still beating and seems relatively healthy is important to note. This is especially interesting given the severely infected lung tissue. Healthy lungs bring in oxygen for transfer to blood that is then pumped throughout the body by the heart. Severe pneumonia ought to interfere with oxygen transfer. In fact, we have long wondered why ANSD victims do not succumb to pneumonia early in the process. Blum's theory, that something protects the process of respiration, is a major theoretical development. We are pursuing this theory now on tissue specimens at the UN laboratories here.

What remains very clear is that Blum and his team need to move quickly exactly as their health is rapidly failing. To this end, we have read and urge all of you to scrutinize the following material for findings that Blum's haste might have overlooked. Could there be observations in his notes—regarding the skin, the muscles, other features of the subjects' anatomy, for example—containing important clues that he missed? Is there something relevant and new in his observations? Would immune responses that appear to protect brain structures also protect vital organs such as the heart and the lungs? Are there any infectious agents that might create immune barriers in both brain and thoracic regions? Blum has only recently been even thinking about these kinds of issues. Please keep in mind that he is not an accomplished scientist or researcher. These questions are in fact central to the focus of tomorrow's meeting.

9:17 AM

Climate control is again intermittent—I'll need to stop at midday to avoid the heat.

Before examining the lungs, I decided to culture some of the skin ulcerations. The wrists seemed most secure, so I scraped the skin off the side of the subject's hand onto a tongue depressor, the dermal layers peeling away like onion skin, and pasted what I could onto the few petri dishes that aren't already contaminated. If there is a novel pathogen, if something is protecting the lungs and the heart, then there has to be an entry point. Maybe some of the infection comes from direct contact...Maybe the skin lesions themselves provide the opening.

The subject woke suddenly during the skin cultures, snapping its jaws, again seemingly oblivious to pain. Breath is foul and rotten. I let the skin flaps fall off the face. Heart is visible beneath the ribs, and the pericardium is boggy...leaky. Blood everywhere. Whenever I move, the heart rate increases. It knows I'm here.

I suctioned off the excess fluid to better visualize the heart. Using bone clippers, I bisected the sternum and lifted the rib cage. Bones mostly crumbled. Heart muscles are hypertrophied, probably due to increased

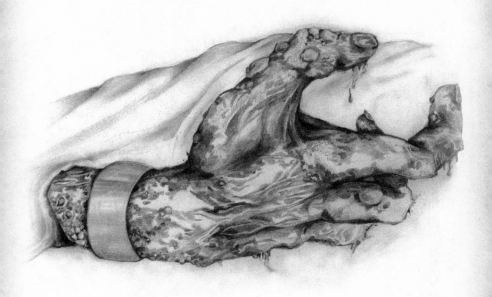

Note the presence of external infection and friable
tissue visible along the skin of the restrained hand.

Note the enlarged heart and decaying rib cage.
Subject seems unaware of surgical status.

metabolic demand from fever and dehydration. Pulse is 145. Pericardial fluid was drained and collected.

The heart is impressively preserved—it looks healthy, again as if something is protecting it from the invading pathogens. The contrasts...The rib cage is meant to protect the heart and the lungs, and yet the ribs crumble with almost no pressure. The heart, however, is tough, robust. Every time it beats, it literally moves the structures around it. And because of the fever, it is beating very fast, well over 140 times per minute. I don't know if we appreciated this contrast in the initial studies. I certainly didn't note it in my reading.

It is almost as if there is an evolutionary process here. The pathogen changes the host, taking it out of niches that it has inhabited for hundreds of thousands of years. It turns us into insects, into bacteria, into primitive things. It broadens our biological niche at the expense of everything that makes us human. It confers to the host the ability to survive multiple attacks, multiple onslaughts. These things don't die. It's like trying to kill a cockroach. The heart is healthy, even robust. Cardiac muscle is enhanced by hypermetabolic states consistent with infection. Higher brain function is no longer necessary and is thus degraded, destroyed. Cortical material might even constitute nutritional value to the ANSD pathogen itself. Humanoids aren't the hosts. They're the disease itself. Stage IV ANSD victims become the disease.

We don't hesitate to kill cockroaches. We step on them. For bacteria, we give antibiotics. We do not suffer a contagion to live.

Maybe we've got this all wrong. We treat strep throat with antibiotics. But "treatment" isn't the same as "eradication." If the host is the disease itself, then we have to eradicate the host.

And we are all, all of us, infected. We're all hosts.

I can't get my mind around this. These things, these cockroaches, have human hearts that beat. They're alive. How can they not be alive? What the hell does NLH mean? <u>What does "No Longer Human" mean?</u>

It knows I'm here. It is sentient, aware.

But so is a cockroach.

And cockroaches can be killed. Cockroaches can be drugged and crippled and baited. Unfortunately, there is also a long history of quick and life-preserving adaptations among primitive organisms. Cockroaches have mutated, have become resistant to many of our earlier poisons.

Remember when we tried warfarin, rat

poisoning, as a means of slowing or killing Stage IV subjects? They remained animated much longer than expected. They bled more but they kept moving. And the disease remained viable, still transmittable even after the heart stopped beating, even after all that bleeding.

I don't think we'll cure this without vaccination and treatment combined. We need to stop future generations from getting the bug in the first place, and since we are all already infected, we need whatever vaccine and treatments we develop to prevent disease progression. We need a regimen that both fights and eradicates the disease. We need to take back our immune system, get it to work for us and not against us. This is an interspecies battle.

10:15 AM

I had to leave. Had to see the sky, the ocean.

But the sky is green. The sky, the ocean, the air. All green. All wrong.

There were tentacles washed up on shore, disembodied from the thorax of whatever they once occupied. A squid? Octopus? Some cephalopod without the skull to stop its brain from exploding. SCE. We don't see much of that anymore. Most are gone, already eaten by whatever is lucky enough to still be alive.

I kicked it back out to sea.

That thing in the lab is alive. More alive than any of the videos they showed us.

And it's hungry.

Actually, it's starving. It is tied to the operating table—I tied it there—and it is starving.

That thing is dying.

The wind shifted and the remaining humanoids picked up my scent. They started howling, moaning, and the one in the lab moaned, answered. Do they miss her—miss it? Does she miss them?

We nuked them and they survived. The humanoids died but they survived. The bug lived. Like a cockroach.

We've tried anti-viral treatments but it survived. The host looked worse, but the virus survived.

Remember the initial treatment trials in Atlanta? Bethesda? Geneva? Humanoids looked worse with every anti-microbial measure. Humanoids need the pathogen. If we kill the pathogen, we kill the host.

But the host is already dead. That's what they told us in Atlanta.

"They're gone."

The robust heart, the functional hypothalamus,

the reticular activating system. All these structures keep working. Our organs keep working, because the virus <u>makes</u> them work. We get hijacked. We become the virus.

The green sky? Doesn't matter. The dead fish? No impact. Viruses don't care about beauty. They don't care about art. They don't fall in love.

Mitochondria—the oxidative factories for life. Don't the evolutionary theories suggest that mitochondria once lived on their own? Things got too rough for them, and they used our cells for shelter. They used higher organismal cells to reproduce, and in exchange they created energy for those cells. That's why oxidative metabolism is so robust. It's an evolutionary exchange that fills a symbiotic niche.

Gupta's heart...I've never seen a heart look so healthy in something so sick. The pathogen keeps it beating. It uses it. Uses us like mitochondria.

Jesus, I'm losing it. Feel dizzy.

If the host and the virus are the same, then something makes that possible. Some immunological response allows—no, <u>requires</u>—the change. Stage IV ANSD is characterized by the "loss of all that is human."

Turn that on its head. Stage IV is characterized by the arrival of something new. A new species...

What gives it that power?
I need to hurry.
What else is preserved? What else keeps working?

11:12 AM

I woke Pittman. He looks shaken, won't say much. We need to work. That's all I said, and he followed. He sat on a chair and drew pictures, like he was in a trance, like he was just going through the motions. But when I checked, he was drawing what we needed. We need to document this, I said, and he nodded, not looking up. His sketches of humanoid lungs emerged on the paper, the pathology oddly sterile in the absence of color. But he can still draw.

Alveoli are almost completely obscured. How do they breathe? Looks like ARDS—Acute Respiratory Distress Syndrome—a condition in which one or both lungs appear completely occluded by inflammatory response to disease and thus cannot engage in respiration, in the utilization of oxygen. However, unlike ARDS, these lungs have chronic rather than acute changes.

How do the lungs move oxygen? What carries the oxygen? Is something other than the lungs allowing more efficient oxygen exchange? There's no way there's any oxygen transfer in these lungs, in this rotten tissue. And the oxygen-carrying capacity of humanoid red blood cells is almost zero. Again, there's got to be some compensating mechanism, some process that we haven't considered.

Because the ANSD pathogen moves freely throughout the host, is it possible that the pathogen itself has evolved an independent oxygen-carrying capability? To my knowledge, no new oxygen-binding compounds have yet been discovered in any molecular analysis of the putative virus, but we haven't really looked.

I removed the lungs, weighed them, sectioned the tissue for later analysis. Christ, the lungs are barely even there. They weigh almost nothing. How do these things take a deep enough breath to muster any sound at all, let alone those inhuman howls?

I should look at the peripheral vasculature. If something is carrying oxygen, then the place to find clues is in the extremities, the farthest distance that oxygen would need to travel. Remember, these things can still grab and hold tight. Somehow the arms are getting fed.

I had started this dissection, dissecting the muscle, the veins, and Pittman just fell, collapsed onto the ground. His drawing almost fell into the

Muscles are atrophied, but...

...rudimentary motor control remains present.

muck. I rescued the drawing first, felt bad about that, but it's what I did. Then I woke Pittman. He looked like he'd seen a ghost. His eyes were pleading. "Have to get out of here." That's what he kept saying.

I helped him up—he could walk, but not well. We went outside, into this mist, and then back to the infirmary. There was Gutierrez, still sleeping.

I let Pittman rest and studied his sketch of the dissected extremities. I had exposed the bones, muscles, and nerves, and it was impressive how much the muscles themselves were atrophied. Infection is everywhere. I wish I could talk to Gutierrez. Tried to wake her, but she wouldn't rouse.

The muscles in the arms are shrunken. A robust heart, but emaciated muscle everywhere else. All the nutritional input must go to the heart, to whatever keeps them breathing, to whatever is left of the brain. All the other infections—the fungi, the bacteria, whatever other viruses are there—none of that matters. Something stops all the other stuff from spreading. That's why the arms look so frail and the heart looks so good. That's why the hypothalamus is solid and the cortex is gone.

The arms looked like twigs on a sick tree. Sticks that barely move. The integrity of the nerves, the vasculature, even the bones, all just enough to allow function. Nothing special. Humanoids can't draw like I'm drawing. But they can grab, they can claw. They can move flesh to the mouth, and that's all they need to do.

All of this suggests an almost conscious process. The disease has an economy to it. It preserves exactly what it needs to spread.

Because, all things considered, it is still a virus. It thinks like a virus. Efficient. Dispassionate.

Viruses co-opt their hosts—they even use our nucleic acid, the DNA and RNA of our own cells, to reproduce. A successful virus will cut its losses where it can if these losses favor viral propagation. Anything that isn't needed in the host becomes food for the nutritional needs of the virus. Peripheral muscles break down to their barest essentials. Just enough tissue to allow the arms to feed, to bring food to the mouth, and nothing more. The arms waste, but the temporal muscles, the masseters, the muscles of biting and chewing—these are as healthy as the heart.

They exist only to exist, only to feed.

12:35 PM

I dreamed about funerals. So peaceful, the slow parade of cars down rainy streets. The ritual, the understanding that those who are with us will celebrate and remember those who aren't. Comforting. Necessary. Need more sleep; more comfort.

Forensic psychologists and psychiatrists are confident that Blum's wandering train of thought, his dreams that sound almost like visions, are potentially the first signs of mental deterioration. However, analysts are undecided as to whether this represents early neurologic involvement of ANSD or trauma from the gravity of his experiences. Both are possible. Although he does not demonstrate the classical signs of ANSD shock syndrome—he is very much aware of his surroundings and is not at all dissociated from his consciousness (see appendix I for a complete description of ANSD shock)—we are cognizant that the presentation of neuropsychiatric distress is heterogeneous and multifaceted. *From this point forward, readers are advised to consider Blum's thoughts as neither clearly nor consistently coherent or objective.* Certainly his later entries suggest a rapid cognitive and emotional deterioration, though again this may not be related to ANSD as much as to his own particular temperament and the prevailing setting.

Finally, it is important that we note Blum's observations about the environmental state of the world. He encounters the corpse of a cephalopod, presumably a victim itself of ANSD infection. (See appendix I for a description of Spontaneous Cephalopod Cranial Explosion [SCE].) Different nations have managed the shock associated with the realization of the destructive environmental changes on Earth in vastly different ways. Many in the current UN bunkers have opted not to view available surveillance of outside environmental conditions since the time that nuclear and non-nuclear eradication efforts started. Some nations with citizens sequestered in underground bunkers have in fact forbidden citizens from ever viewing this destruction. We respect these choices, but for those individuals for whom Blum's descriptions are new, please understand that his descriptions are *not signs of worsening delirium.* The green-tinged sky to which Blum refers is the result of both nuclear ash as well as exfoliated flora caused by neutron radiation. Again, to honor the wishes of those countries that have forbidden access for their citizens to view these changes, we request that these descriptions be kept confidential.

NOVEMBER 21, 2012

5:17 AM

Pittman woke me. He had applied fresh bandages to his wound and checked on Gutierrez. He looks better. Calmer.

I told him what I had been thinking about—that the virus itself confers protection to the host. That the host should be dead, but isn't. It is dead, he told me. That's why we call them zombies. He smiled nervously.

But that's not the point, I told him. The point is that the human is dead, but the virus changes us, uses us. We keep moving even when we're "technically" dead.

Pittman didn't like this, and I stopped explaining. They're dead, he kept saying, mumbling it over and over again under his breath. I dropped it. I need him to draw. Need him to hold it together.

We need to get back to the lab, I said. He agreed.

We need to look for molecular signatures of something that we may have missed—something that might be protecting the vital organs from infection.

We need genetic footprints, electron micrographs. We need to look much more closely, section the tissue we've preserved and examine it with immunofluorescence. I'm still thinking of a third pathogen. Nothing that I've read about would allow either influenza or prions to create these kinds of protections.

Maybe we should focus on organic structures that are preserved despite the infection. Why doesn't the hypothalamus degrade? Why are the amygdalae intact?

And what about the GI tract? Humanoids eat. They change "food" into energy, and that means that whatever they ingest of nutritional value must be absorbed in the stomach and bowels. However, past dissections have shown that the bowels are virtually dysfunctional. What did we miss, then? It has to be at the microscopic level.

I need Gutierrez for this work. I can't even begin to do these analyses, but she needs rest.

Pittman and I agreed to go to the lab, to prepare the tissue samples we've saved. We'll start by sectioning the hypothalamus at the same time that we open the abdomen. We'll get things ready and then one of us will awaken Gutierrez. Let her sleep until we need her. We'll move on to whatever tissue she wants to see next.

6:35 AM

I had forgotten about last night, about what I had done. When Pittman and I got to the lab, he asked me where the subject was, and at first I didn't remember.

Pittman looked around the room, frightened. Is it loose? he kept asking. Is it free?

He was holding a scalpel, shaking. I tried to recall what happened last night.

I burned it. It was still animate, but last night I bound its arms and stuffed gauze into its mouth and placed the humanoid onto a corpse-size slab of plastic that had washed up onto the rocks. I wrapped rope around the body, fastened it to the plastic, and set it ablaze. I used the flamethrower from the arsenal, and while she was still burning I pushed Gupta out to sea.

I believe Hindus cremate their dead.

For a while, the ocean lit up like a torch, but something was still alive in the ocean. It bumped the pyre from below, tipped the body into the water.

The fire went out.

I didn't tell Pittman, though…couldn't. Told him that the subject deanimated, died, that I threw it in the ocean to avoid more crabs.

They don't die that easily, he mumbled.

He was angry. Wanted to know why I didn't feed it to the two subjects in holding.

I didn't answer.

Pittman stared. It bit me, he said. Goddamn thing bit me.

We left the lab to get Gutierrez, but she wasn't where we'd left her.

She was in a wheelchair, in the kitchen. We ate beef jerky, drank warm, only partially desalinated water. No one spoke for a while.

I mentioned the plan to study the digestive tract, that we need to know what they do with what they eat. We need to understand why they don't gain weight. What they're running on. Gutierrez agreed, and she was particularly interested in the nerve cells of the large and small bowels. Again, she refused to say more, except that she needed genetic footprints of any virus in the colonic neurons. I tried to get her away from Pittman so that I could tell her what I was thinking about the virus, about its immune properties. Couldn't, though. Pittman stuck to us like a frightened child. I can't risk losing Pittman.

6:40 AM

The holding facility is foul, wretched. Threw a crab and a living fish into the pit, and the one that could walk got both, gumming the creatures with almost frantic intensity. The other had lost strength, coordination. It dragged itself toward the food but didn't get any, rolled over onto its back, and moaned.

Pittman raised the platform and grabbed a transport pole. He reached over the fence and pinned the one that couldn't walk onto the ground and shocked it. Shocked it over and over, its toes pointing down. Gutierrez shouted at him, told him to stop, that we couldn't afford to lose a specimen. Pittman just stood there, holding it in place. It stared at him, its eyes glazed from the voltage, and I didn't have the strength to intervene. Pittman finally let go. The subject was still breathing.

What's happening to us?

We took the healthier humanoid to the lab, removing the old shackles and locking on fresh ones. The voltage from the transport pole held it still while I locked the chains. There were mites and snails feeding on its calves.

No difficulties moving it, prepping it. It woke entirely just as the head was secured to the table. It growled.

The abdomen is bloated, infected. The surface is hard, almost wooden. Peritonitis. There must be bacteria throughout the peritoneal space. The entire abdominal cavity must be colonized by bowel flora. Nothing survives this. Nothing human.

I made a vertical incision from the base of the sternum to the umbilicus, and then laterally, down either side of the subject's lower abdomen. The dermal tissue and muscle peeled open, like an antique book. The smell was overwhelming. It's been more than thirty years since I smelled dead bowel, but no medical student forgets this. It smells like death.

The spleen is huge, but I guess this makes sense. Makes it more human. It's like the body is taking sides. The cortex surrendered but not the goddamn spleen. Fucking spleen. It's pretty worn down by now. I tried to remove it but it popped, like a balloon.

Stored what I could of the spleen, liver, and gallbladder in some of the remaining preservatives.

I looked across the room and saw that Gutierrez was stooped over her workstation. She was sectioning the hypothalamus from a few days ago, preparing samples for microscopic analysis. We had sectioned the hypothalamus into distinct subanatomical structures. Different specimens were spread across a lab table, each sitting on top of a note

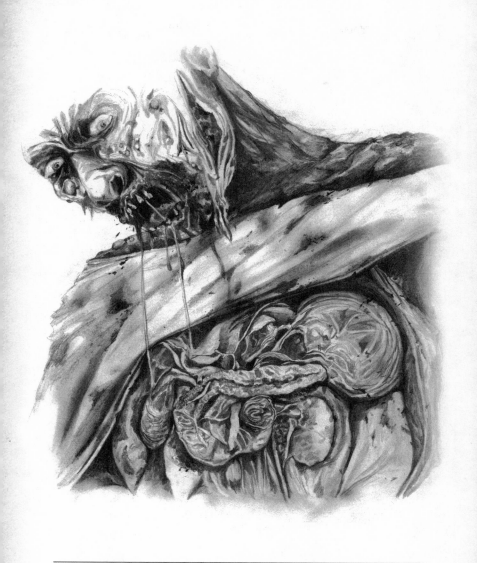

Spleen is enlarged and displaced.

card with writing. There were descriptions, shakily written by Blanca.

The "anterior hypothalamus" sat on one card, the "posterior" region on another. And so on...

Gutierrez was smiling and called me to the computer. One of her samples was under the microscope, and its image was projected onto the monitor.

Look at this, she said.

I didn't see anything. Almost no cells at all. Is the stain OK? What's wrong with the sample? She didn't say anything, but changed slides, told me to look again.

This time the histology, the cellular architecture, looked fine. Better than fine. It was intact. Microscopic boundaries were distinct. There was, however, a strange density of mononuclear immune cells. Immune cells devoted to fighting something, but mononuclear cells do not typically attack viruses. At least they aren't the first-line immune response for most viruses. This was healthy, robust, hypothalamic tissue that was, nevertheless, compromised by something. Some outside pathogen called those mononuclear cells to attack, and the tissue looked different from what we had seen in ANSD infection.

She showed me the note card from the sample I

was looking at, the one with the mononuclear cells. It was the "ventromedial hypothalamus." She pointed to the note card for the previous slide. "Posterior hypothalamus."

The ventromedial hypothalamus is relatively healthy, infected by one pathogen probably, given the mononuclear cells, but the rest of the hypothalamus is junk. ANSD won't attack the ventromedial region, she said, or else something that already lives there, some other pathogen, is fighting it off.

Another pathogen? I asked.

She smiled.

11:34 AM

Autopsy going well. Body block in place, incisions made with little difficulty. Bones and fine tissue in worse condition than other subjects. Humanoid remains animate.

I saved its dog tag, hiding it in my notebook. Wonder who it was. Haven't read the name yet.

We still need to understand how they digest food. Previous studies have shown that the gut is in ruins. Perforations, peritonitis, infection. How is nutritional material absorbed? Presumably everything humanoids eat feeds the virus. If we can find a way to interrupt this process, maybe we can starve them long enough for a vaccine to take hold.

The abdominal cavity was infested with parasitic worms, many more than three meters in length. Gutierrez was working on identifying them.

And the bowels are definitely malabsorptive. Fecal material is scattered throughout the peritoneal cavity. Multiple colonic punctures throughout the GI tract; undigested crab shells clearly identifiable.

Again, a human would have died from this kind of infection, from the effects of fecal bacteria and parasitic worms literally swimming in the abdominal cavity. The belly is rock hard—a first-year medical student would recognize this as fatal peritonitis.

Other powerful viruses—Ebola, for example—quickly kill their hosts. Usually isolation of infected populations allows the disease to eradicate its own spread. The host dies and there's no one alive to carry the virus forward.

This bug is different.

Blanca reminded me that temperature is regulated in the hypothalamus, and that humanoids can still mount a fever. I think she's noting that they ought to be too sick to do this, to even get a fever in the first place, unless, once again, something were protecting the hypothalamic response. She also reminded me that the hypothalamus is where hunger sits, that the subject is hungry and feeble but still febrile, still wanting.

I watched the subject, its eyes wide and seemingly unaware that we were removing its colon. There is something grotesquely admirable here. A life force that feels old, almost ancient. Something primitive.

Forgot to ask her about the goddamn goats. She had drawings of goats in her office, and I don't have the slightest idea why...

Keep forgetting things.

Sun is hot, and air-conditioning is out. We're going back to the bunkers. Need sleep. Need to think about this. Gutierrez to stay behind; need time to culture the sections. I didn't want to leave her there alone, but she insisted. Too tired to argue.

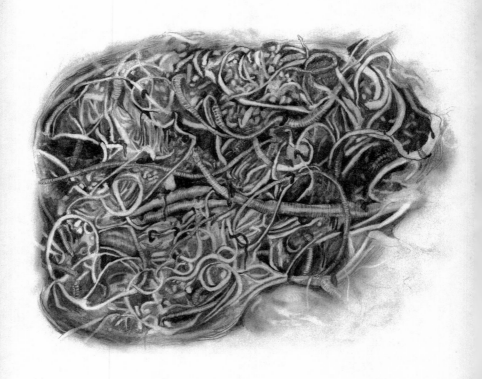

Opened abdominal cavity with parasitic worms and bowel perforations.

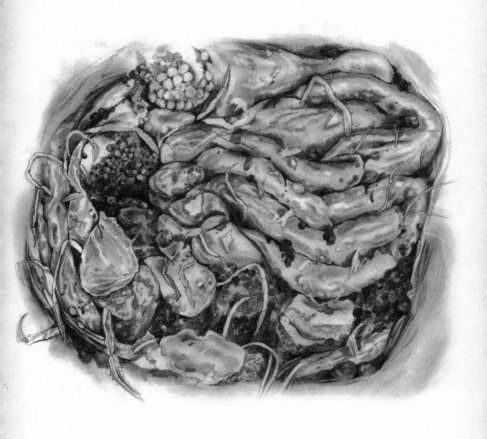

Small bowel removed showing ongoing presence of tapeworms throughout open cavity.

At this point, we are confident that Gutierrez feels she has at least isolated clues to an additional pathogen. We do not have a good understanding of how this pathogen would lead to the enhanced ability to stay animate. As Blum notes, a vaccine will need to recapture the immune system of the host. Because neurological damage and the resultant behavioral changes appear irreversible at Stage IV, we'll need to study subjects at earlier stages of disease progression. This means dissecting humans at Stage III or even II of the disease. In other words, we must consider dissecting living humans, a clear violation of the Atlanta Protocol. Please arrive tomorrow prepared to discuss the ramifications of this possibility.

NOVEMBER 22, 2012

10:11 AM

We slept for more than fourteen hours. Ate nothing last night.

Gutierrez was asleep in the lab, lying in the fetal position on the ground next to a lab stool. She must have fallen. When Pittman and I opened the door, she woke, looked confused, even bewildered, and then she immediately asked for sections of the large and small bowel. We helped her, pushed the stool back to the table. She seemed excited, moved quicker, but kept dropping things. She perspired onto her notes, onto her workbench. Asked us to bring her a towel.

She passed out twice.

The first time she fainted, her head hit the corner of the table. Blood ran down the side of her face. The force woke her and she motioned to her pocket. We found smelling salts there; she'd been hoarding them. Use them, she said, and we broke them and waved them under her face. Did this the second time also, and both times she pulled herself onto the stool without saying anything. She was breathing rapidly. Pittman brought an IV pole for her and I somehow found a vein for the needle. We kept her hydrated.

She asked again for sections of the subject's

bowel. I had forgotten the subject was there. Its heart was still beating, leaking all over the table. It was awake. The bowel was tangled and perforated. No clear demarcations from large and small bowel. Lots of fecal material. Parasitic worms hung from the tissue samples. Whatever food is absorbed goes beyond what these parasites can divert. Perhaps this adds to the hunger. Subjects need to eat nearly constantly in order to absorb any real nutritional content.

She explained to me that she needed the sections no more than five microns thick. Not enough power left for the electron microscope. She was out of breath after telling me this, how to do this. I stood at the other lab station, trying to hold my hands steady.

I'm not sure how to explain what happened next but it is important that I keep a record. Future studies might take place. There may be others who continue our work. We need to be aware of the ways psychological distress can explode. Later I'll try to analyze these events, try to see if I could have predicted them. Right now, I need to note them, need to let others know about what happens in these circumstances.

I noticed Pittman first. Heard him, I guess. I was so focused on preparing the tissue that I didn't see him get up from his stool. He was whimpering, kept trying to draw. He had a pen, a large pad. Kept tearing out paper, throwing it at the humanoid. Can't get it right, he said. Can't stop shaking.

Ignore him, Gutierrez whispered. Keep working.

I listened to her and went back to work, but I should have helped Pittman. Should have paid attention.

Pittman pushed the buttons, undid the restraints. He set the subject free.

He was crying. His crumpled drawings were stuck to the zombie, sinking into the half-empty abdominal chamber. It rolled off the table and crawled toward Pittman, its eyes glazed. It was growling, tracking our movements.

Pittman threw himself at it, stomped on its legs, inside its abdomen. He slipped, fell, his face landed on the subject's mouth. It bit his tongue, wouldn't let go. I reached for a tranquilizer and shot him. Meant to hit the zombie, but got Pittman instead. He slumped on top of it, the subject closing its arms around him, thrashing.

Gutierrez fell from her chair, reaching for a gun. I kicked Pittman off the subject and shot it twice, the second dart hitting its open mouth, disappearing inside.

It stopped thrashing and was still. It was deanimated. The dart must have hit the brain stem.

Gutierrez pulled herself up again and told me to forget about Pittman, that time was short. We only have one more subject. Finish the section, she said. I need it. Need it now. Then bring me the other subject.

I stared at her but she had already forgotten I was there. She had rigors, a fever. I could feel her fever from across the room. She was trying to get a slide under the microscope.

We can't sacrifice our humanity. We can't. I put Pittman over my shoulder and carried him to the infirmary. Blood poured from his mouth and soaked my shirt. How the fuck do you bandage a tongue?

Tea.

Can you believe it? All this technology, all this crap, and the first-aid kit tells me to use tea bags to stop the bleeding. I left Pittman on a stretcher and ran to the kitchen, returned a minute later and stuffed his mouth with gauze and Earl Grey. That was all we had left. Gupta preferred English Breakfast.

And then I started laughing, looking at Pittman, unconscious, caffeinated, gauze and tea bags hanging out of his mouth. Shoved a needle into his thigh, gave him some antibiotics. Couldn't stop laughing. Strapped him to the table and don't even know why. Didn't trust him. Don't know who's crazier.

I was leaving and then turned, went back into the infirmary. Pittman's face was red. His eyes were closed, but he was hot. Burning. He was mounting a fever. Something in the bite on his tongue was already attacking him, already mounting an immune response. There must have been thousands of bacteria in the mouth of that zombie.

It dawned on me slowly, like I was drunk and just realizing it. Infection, non-ANSD infection, triggers an immune response. The human immune response lowers pH. Pittman was going to change...

I undid the restraints and lifted his arms into the air—they stayed stiff, shaking. A seizure. He was biting the tea bag over and over, his fever worsening, his respiration rate increasing. He's going to die, I thought. He's going to change. How many organisms were in that zombie's mouth? How many pathogens? He can't beat this.

I sat on the chair and watched him, his whole body twitching with the seizure. He had ANSD neurological involvement. Late Stage II, closer to Stage III. The bite had triggered accelerated progression of the disease. His immune response was engaged; couldn't have a fever without that, but it was helping ANSD to spread even as it was fighting the newly introduced bacteria from the bite.

Acidity. Infection. Immunocompetency. All of that creates an acidic environment. His pH must be dropping, plummeting. I yanked up his shirt and checked his pH meter. The dial was moving from purple to red. I could picture the prions in his brain folding, changing. I could picture them moving.

I gave him more antibiotics to quell his own immune response. I want the antibiotics to kill the bugs infiltrating his tongue. If the antibiotics can kill the bacteria, then his own immune response will quiet down and his pH will stabilize.

No change—still seizing. Still changing.

I grabbed some bicarb, injected it into his thigh. I attached the quick leads for the EKG. Watched the needle on the meter move higher, better. His pH was rising. More bicarb, more Lasix.

The alarm for his EKG rang—he had U waves with ST depression. His heart was faltering as his pH increased. Alkalosis affects electrolytes. Lowering pH decreases intracellular potassium, and without potassium in the cells, the heart can't function. That's what the EKG was showing. His heart was failing.

I crossed my fingers, and watched. Had no idea where potassium replacement was. I felt like I was watching him die.

Then his EKG normalized and his pH leveled out. Somehow his heart overcame the rapid alkalotic process. But that meant the prions were moving again, twisting, changing.

Gutierrez on the intercom. Told me to come back to the lab. That she had something to show me. I stared for a second at Pittman, and then left. Didn't restrain him...Just left.

NOVEMBER 23, 2012

1:00 AM

Woke up on the floor of the infirmary. My head was bruised, Pittman breathing normally, most of the gauze out of his mouth.

I had fainted. Moved too quickly. Dark outside. Pittman still on the bed.

I remembered that Gutierrez had called me. Looked at my watch—she called more than two hours ago. What had she found? She had started to analyze the colonic neurons when I carried Pittman to the lab. Wouldn't they be related to the neurons in the hypothalamus? Doesn't the gut communicate with the brain through colonic and hypothalamic neurons?

I tried to picture her theory as I stumbled toward the lab. She must be thinking that the same organism that targets the hypothalamus also targets the gut. That must have been what she wanted to show me. She might have isolated her pathogen.

I arrived at the lab and found Gutierrez. Sort of. Humanoids don't understand wheelchairs.

She was on the ground, pulling herself up to a window. She growled when she saw me, lunged.

I put a chair between us. She seemed weak, couldn't move it. She pushed it from side to side, the chair rocking on its back legs. On the stainless-steel gurney were scribbled notes next to small slices of hypothalamus. The notes were held in place by a pile of bones and shells. The pile was balanced, neatly stacked. It looked like one of those Zen sculptures. She had drawn a sheep using humanoid blood, a stick figure on the wall that I wouldn't have seen if she hadn't pushed the table just under it. It looked like a cave drawing. I was supposed to see these notes and the drawings; she meant for me to see them.

I looked around the room. The microscope monitors were protected behind a file cabinet. She had moved them so that they wouldn't be harmed. Slides were in place and an image was projected onto the screen. Mononuclear cells and healthy tissue. The tissue was neuronal, but it looked different from the hypothalamic sections from earlier. I looked at the microscope. Using surgical tape speckled with blood, she had attached the note card from the microscopic specimen next to the sample. It was hard to make out, but I could read it. "Colonic neurons." I looked back at the monitor.

It looked so much like the infected ventromedial hypothalamus. Same process. Same architecture. I was mesmerized. There was something here! She knew it, was on to something. Something invades the hypothalamus and the colon—provokes the same immune response in both regions. The same bug. But not ANSD. Something new. Something we hadn't considered with the original vaccines.

I tried to study the stick figures, the sketches on the wall.

But Gutierrez kept lunging at me. All I needed to do was to kick her, move her out the way. She was a skeleton, barely even there. I didn't even have to kill her.

But I couldn't do it.

Instead I sat down on the other autopsy table

and watched her. Watched it. She was No Longer Human. She was gone. I looked around again, noting the absurdity of my sudden gift of time. She couldn't get me. I was a clown, safe behind a steel chair and table.

The lab was destroyed, instruments everywhere. Windows broken, crabs feeding.

The subject was half eaten, a carcass on the ground. The ribs were missing, now part of the Zen pile across the room.

I wanted a funeral. Was she Catholic? What do Catholics do when they die?

I started thinking about the notes, and walked over the tops of the tables to get closer to what she had left me. She watched me, dead eyes, growling. She pulled herself along the ground and was able to grab my leg. I kicked her face and she flipped over, onto her back, her tongue wildly thrusting in and out of her mouth. A reflex, I told myself. Her tongue— only a reflex.

Her notes were hard to make out. Lines of writing were all over the lab paper. There were no complete sentences. I wrote down all that I could read. "A new pathogen...No current vaccine" was scribbled off to the side. She had sketched more animals, more livestock, written something below the childish sketches. It looked like "Born": Something new was born? What did she mean? What was born? Was she thinking about our species changing? A new species is born? She was still staring at the ceiling, licking her lips like an insect, like a fly cleans its face. My breathing caught her attention and she began moaning, dragging herself toward me again. I climbed on top of a table and watched.

And then Pittman arrived, staggering, carrying an IV pole. Look! I shouted, and pointed at the wall. What does she mean? I wanted to show him the notes, all that she had left us.

He swung the IV pole toward me and missed, knocking a transport pole onto the ground. He was using a tool...zombies don't use tools. He wasn't gone yet. Stop, I yelled. Help me!

He staggered closer, his swollen, bleeding tongue hanging out of his mouth. He looked drunk, ghoulish...

But he was aware, still alive. Still human. He tried to talk but couldn't, instead making a muffled, thwarted sound. His tongue.

I ducked, thinking he was after me, but he wasn't.

The pole crushed Blanca's temple, broke her cheekbone. He hit her again and again, and she stopped moving, her skull fractured, her brain ruined.

And then, in one fluid movement, he dropped the pole and shot himself. Put a gun in his mouth, pulled the trigger.

Where did he get that gun?

Blood everywhere—Blanca's notes were soaked, the writing smeared with what used to be Pittman's brain. I could still make out some words, though. And there was that drawing of a sheep on the wall. A child's drawing.

I grabbed a pen and scribbled all that I could decipher of Blanca's notes. "Livestock contaminant...animal to human jump." A drawing of a cow, maybe a horse. No—that's a cat. Horses don't have whiskers like that. "Sweden, 1995, staggering disease." What did that mean? "2005—Communicable Disease Unit. UK. Farmers, agriculture...assoc. with obesity, eating... parrots. pets."

2:15 AM

I've been sitting here, trying to focus, trying to think. At first I kept turning Blanca's notes over and over in my hands, but the writing was disappearing, mixing with the blood and tissue. It was blurring.

Some new pathogen. Something new. Something

130

we hadn't accounted for. Something we hadn't protected against in all of the vaccines.

Or was this the same organism as the ANSD pathogen? There's no way we know everything about this bug... What if the ANSD virus has another organism combined, intermixed, symbiotic?

The odds of that happening in nature are pretty slim... But we always thought someone made this... that it didn't happen on its own.

The influenza part of ANSD makes it contagious, makes it spread. The prions change the brain, create rage, aggression. No higher cortical input.

But what about the hunger? What makes it hungry? We thought initially that a primitive brain was hungry by definition. But what if ANSD humanoids are hungry because something makes them hungry? Makes their brains hungry. Some infectious agent makes them hungry?

So, two separate pathogens in the same host? Or something new? Something with distinct and adaptive properties. Something that hijacks the host. Something that changes us.

I was trying to think, but it hurt. Everything hurt. My ears rang from the sound of Pittman's gun. My head feels like it might explode.

I was still standing on the steel table, but there

was no reason for this now. I stepped down carefully, tried not to look at Pittman or Blanca, and I slid into a chair.

5:36 AM

I woke because of silence.
All I could hear were crabs and insects.
No moaning.
The last subject!
There ought to be one left, still in the holding area. It would have heard all this. If it were still animate, I could study it—try to repeat Blanca's work. Even if it had stopped moving, had died, I could still dissect it, could at least try to confirm what little sense I could make of Blanca's work.

I moved as quickly as I could to the holding area, stumbling, the ground seeming to move with the waves of the ocean, the humidity like a wall, the flickering lights of the lab directing me like beacons.

The gate to the facility was open. Pittman must have been here. There was a transport pole around the last subject's head, the pole leaning against the wall, the power in the pole empty. It had been jimmied, surgical tape wrapped around the handle and holding the button down. One long, continuous shock. The subject was deanimated, its brain burned, smoking. There was a smell. I looked at the body. Devoured, crabs feasting. There was no point to dissection. Nothing left.

23 November 2012

(DATE ADDED BY UN ANNOTATION TEAM BASED ON FORENSIC ESTIMATIONS)

4:35 PM

woke suddenly. Head crushing. My head. So much pain.

pH meter cracked. No reading.

Sweating.

Hungry.

I haven't felt hungry in weeks, maybe never really felt hungry. Not hungry like this. I can smell the crabs. Smell the bugs.

Laughing again, blood dripping from my nose, my mouth, everywhere. All over these pages.

I read back, read I wrote. Remembered. Sheep. Blanca. both dead.

I'm changing. Hungry.

I can't walk, can't hold the pen. I am legible?

sheep drawn in blood

notes. where

Laughing again. Why?

Who is this for? who is so hungry so strong.

To the infirmary. Thought I was walking. Saw my reflection in water.

Not walking.
I feel strong. Feel invincible. Just need food.
Feed. To feed.
The infirmary.
I'm changing.
Changed.
The infirmary
needles. someone
need needles.
needless this end.
blood blood on needles
I'm alive.
Have to eat

Digital imagery from the excavation team did not locate any of Gutierrez's writing. The drawing that Blum describes on the wall is gone, though there are bloodstains where Blum appears to have been looking. These images will be available at tomorrow's meeting.

We view the content of this journal as immensely important and only the beginning of a new line of inquiry. If there is indeed an additional pathogen that has somehow become part of the ANSD infection, then we can alter our vaccines to create a more focused antibody response. Mononuclear cells such as those Gutierrez isolated are not recruited to fight either influenza or prions. This seems to add credence to the possibility that unaccounted-for infective agents are present.

But this means further investigations will have to involve infected individuals who have not yet reached Stage IV. The emotional and ethical ramifications of this reality are daunting but cannot be ignored. The future of our species depends on reading this material carefully and dispassionately.

Please know that we have already dispatched investigatory teams to remaining areas of intact infrastructure throughout the globe. If the creators of this plague are alive, we will find them. If they have perished, we will find what is left of them.

One-third of humanity lives. One-third of humanity insists that we not lose hope. But all of humanity is probably infected. There isn't much time.

Drs. Blum, Gutierrez, and Pittman will be recognized tomorrow in accordance with their respective families' traditions. Please be prepared to observe these ceremonies before we begin scientific discussion.

Thank you for your time and for your hope.

APPENDIX I

Glossary of Terms

ANSD: Ataxic Neurodegenerative Satiety Deficiency Syndrome. This is the ICD-10 (international classification of disease) term given to the virus that causes zombiism.

- Stage I—Onset of extreme hunger with coexisting fever and upper respiratory symptoms.

- Stage II—Worsening fever with measured temperatures up to 106°F. Cough worsens and cognitive decline begins. Hunger intensifies, with a preference for large, living moving organisms. Balance begins to suffer, with a wide-based, staggered gait. Arms are often held in front to maintain posture. Stage II lasts from one to twenty-four hours.

- Stage III—Significant cognitive and neurological decline. Frequent falls and increased aggressive behavior. Significant malabsorption of food. Stage III lasts no more than four hours.

- Stage IV—Complete loss of human characteristics. Officially categorized as "No Longer Human"—"NLH"—by the UN and the WHO. "Ethically Dead," and subjects are now "humanoids." As death has already philosophically occurred, the cessation of oxidative function among those with Stage IV infection is referred to as "deanimation" and happens after multiple days in the absence of feeding or with the destruction of the brain stem.

ANSD Shock Syndrome: The general descriptive term for the psychological numbing and emotional constriction that many have experienced in the setting of such dire and significant changes to our species and our planet. The condition resembles a dissociative psychological state and is characterized especially by the loss of respect and appreciation for living and viable organisms in the setting of the current crisis.

Gutierrez protocol: The process suggested by Spanish microbiologist Blanca Gutierrez. Gutierrez found that creating a metabolic alkalosis (raising the body's pH) by infusing bicarbonate and administering oral Lasix (a diuretic that causes urination) could prevent the prion-like aspect of the ANSD virus from bending and leading to the neurological changes responsible for the behavior of Stage IV–infected humanoids. However, people with artificially raised pHs often feel quite ill. They cannot think clearly, and almost always die if they do not correct their acid–base balance. Conversely, allowing the pH to correct creates the environment that allows the prions to become infective.

Humanoid: The official WHO-sanctioned term for humans infected with the ANSD virus and showing the neurobiological changes characteristic of Stage IV disease.

NLH: No Longer Human. A specific classification stemming from the Treaty of Atlanta. "NLH" signifies the loss of human status to those manifesting the full ANSD syndrome and is a compromise term for those not comfortable defining infected individuals as no longer alive.

Prions: Infective proteins that have been implicated in the brain changes causing much of the aggressive behavioral changes seen in Stage III and Stage IV ANSD infection. Prions are technically not alive— they lack DNA or RNA—though they behave in many ways like infective organisms and target brain tissue. The epidemic of mad cow disease in the late twentieth and early twenty-first centuries was caused by prions, as are diseases such as Creutzfeldt-Jakob disease (CJD) and Kuru. Scientists believe that prions become infective and toxic when they fold onto themselves. The reasons for this change in their shape leading to increased virulence are still not understood, but it is known that prions are more likely to change to their virulent forms in low-pH (more acidic) environments. This is what led to the Gutierrez protocol.

RAH: Reptilian Aggressive Hunger Syndrome. An early name for ANSD, based on the observation that infected humans behave much like reptiles when starved, utilizing regions of their brain that make up the reptilian nervous system in the absence of any higher cortical function. The WHO eventually changed the name to ANSD in order to better characterize the ataxia (poor balance) and constant hunger (satiety deficiency) also characteristic of infected humans.

SCE: Spontaneous Cephalopod Cranial Explosions. Cephalopods—invertebrates such as octopi, squid, and cuttlefish—all have relatively advanced brains in the absence of protective skulls. Perhaps because of their advanced central nervous systems, these animals are also vulnerable to some aspects of the ANSD pathogens but cannot tolerate the increased intracranial pressure that the condition creates. As cranial pressure increases, many suffer seemingly spontaneous explosions of brain material. The phenomenon was first observed by divers in the Great Barrier Reef of pre-plague Australia, and later throughout the world as previously unrecognized carcasses began washing ashore.

Sony Implant Peritoneal pH Meter and Bicarb Infusion Unit: A device developed by the Sony Corporation to emulate implantable insulin pumps by measuring body pH and allowing the infusion of bicarbonate to maintain the alkalotic state necessary to stop prion virulence. All devices have meters that designate safe, low-risk, high-risk, and pathologic pH readings from the perspective of what acidic environment is necessary for prion activity. They also are equipped with leads to attach to electrocardiogram (EKG) machines. Workers at the Crypt were all outfitted with this device, as are many in UN bunkers.

The Treaty of Atlanta and the Protocols of the Atlanta Treaty (see appendix III): The documents produced by an international meeting of scientists, ethicists, religious leaders, and policy makers that took place in Atlanta in July 2011 at the CDC before Atlanta was overrun by ANSD. The meeting was convened to arrive at a world consensus of how best to characterize infected humans. If people with ANSD were considered human, then ethicists felt strongly that killing ANSD-infected individuals constituted murder. Because the infected were not using any higher cognitive function in their attacks, and because they could relatively easily be outrun and avoided given their cognitive and neurological difficulties, ethicists stressed that one could not reasonably argue self-defense when taking the life of someone infected with ANSD. Similarly, there are no precedents for mass self-defense other than war, and "war" is not an appropriate term without a formal declaration from the enemy. Additionally, if ANSD victims are human, then they are subject to the same caveats that protect all living organisms, especially humans, on which experiments are performed. These issues go to the core of how we define "human" and led world ecumenical leaders to decide from ethical, spiritual, and policy points of view that those infected with late-stage ANSD had, as a result of the infection, lost the fundamental aspects of humanity (the capacity to connect at any kind of human level).

The rapid rate at which ANSD was spreading made it necessary for a quick and cooperative world consensus regarding these issues. The protocols of the treaty dictate that those with the full syndrome of Stage IV ANSD are no longer technically alive, and are referred to as humanoids or, alternatively, as NLH (No Longer Human). Religious leaders approved rituals observing the cessation of life (funerals and so forth) for those infected, and ANSD humanoids are thus considered dead among ecumenical and ethical scholars (though disagreement lingers among this working group).

UNSaSS: United Nations Sanctuary and Study Site. This is a cooperative operation of the UN and the WHO with input from organizations such as the CDC and its international counterparts. The UN/WHO/ ANSD Working Group decided that there needed to be a single location where experiments and tests in the search for a vaccine and treatment for ANSD could take place. Given the quick spread of the virus, there were fewer options than first thought, as an uninhabited and disease-free island was needed. Additionally, the island needed to be close enough to land so that easy transport was possible, and a hot environment was necessary as warmer temperatures increase virulence and make the virus therefore easier to study. A coral atoll, Bassas da India, between Madagascar and mainland Africa was chosen. As it is partially submerged much of the time, no one had tried to escape from the virus there, and the world community funded a quick assembly on the ten-kilometer body of land, using stilts on top of which were placed laboratory equipment, solar power cells, bunkers that can sleep up to ten, a surgical suite, and holding facilities for infected individuals. Due to the virulence of the contagion, every worker sent to the island either contracts Stage IV disease or expires from the persistent metabolic alkalosis necessary to stop progression of the disease. This earned it the nickname "the Crypt" by UN staff. The average life span of humans on the island is about two weeks, and humanoids are no longer transported there; infected humans who are recruited to work at the UNSaSS and who succumb to the disease are instead used as experimental subjects.

APPENDIX II

ATAXIC NEURODEGENERATIVE SATIETY DEFICIENCY SYNDROME

NATURAL HISTORY AND EARLY THERAPEUTIC MANEUVERS*

Mary Anne Fassalini, PhD
Daniel Pittman, PhD, MD
Anita Gupta, MD, PhD
Saori Fujiwawna, MD (deceased humanoid)
Blanca Gutierrez, PhD
World Health Organization Outpost, South Pacific compound
August 2012

INTRODUCTION

Outbreaks of Ataxic Neurodegenerative Satiety Deficiency—ANSD, or "zombiism"—have become increasingly prolific since the initial reports of this disease were first noted among the Aran Islands of west Ireland last year. The history of these initial infections has been well documented (1) but until recently little has been known about the pathophysiology of the syndrome. While a great

*From the international working group on ANSD, South Pacific compound. The authors wish to acknowledge the administrative assistance of the CDC work group on ANSD and especially the advice and direction of Dr. Stanley Blum at the CDC neurodevelopmental branch.

deal remains to be understood, the investigatory endeavors are clearly limited by the increasing scope of the problem and the overarching public health concerns that have required incineration of infected regions. Given these limitations, despite the dire state of affairs the medical community already used a good deal of precious time to study ANSD at a neurobiological, molecular, and indeed political-psychological level.

With these concerns in mind, this paper represents the first attempt at suggesting a cohesive natural history for ANSD based on a comprehensive characterization of the pathophysiology of the syndrome. Much of the data here is based on laboratory work stemming directly from the hastily organized study sites that led to the initiation of the United Nations–designated Sanctuary and Study Site (UNSaSS) for more sophisticated investigations of individuals with ANSD (2). In keeping with the conventions adopted by the Treaty of Atlanta (3), we will not in this article refer to or consider these patients as human. Infected patients are considered "humanoid" once definitive ICD-10 criteria for Stage IV infection are met (4). Given our unique circumstances, we understand and are especially sympathetic with the ethical controversies surrounding this provocative change in definition, but in the interest of stemming the ongoing outbreaks, the reconceptualization of "human" is necessary so that scientific investigations can proceed with necessary inquiries in the absence of more stringent guidelines appropriately governing the use of human beings for experiments. We firmly believe that in addition to increasing the etiologic understanding of the disease and furthering the search for a global remedy, the investigations on Bassas da India will lead to the development of viable primate models for the infection and a therefore more balanced and humanitarian challenge to the somewhat draconian steps taken when the Treaty of Atlanta was drafted a month ago.

The authors are grateful for the world's recognition with the Nobel Peace Prize last month, but also firmly agree with the United Nations Emergency Declaration of early August 2011 stating that the current lack of medical knowledge regarding ANSD is "apocalyptic in scope" (5). The Nobel Prize will therefore be little more than a hollow honor in the absence of a better scientific, medical, philosophical, and environmental understanding of ANSD. The goals of this paper are intended to meet these urgent needs.

BACKGROUND

As noted, ANSD was first reported in the Aran Islands by the Irish Medical Society in 2011 (6). The tourist disaster of 8–15 May 2011 was widely covered by the media throughout the world, though it is now believed that the outbreak began sometime in late April but was hindered in discovery by mistranslated reports from the largely Gaelic-speaking population and the relative remoteness of the islands themselves. What is known with certainty is that the village of Doolan received a distress call on 9 May from a tourist ferry. Irish officials sent to investigate did not return from the rescue mission. The first photographs of the outbreak thus emerged on 10 May, images sent digitally to satellite media outlets. The individuals who took these photos did not survive, and the islands have since been incinerated, forcing medical historians to assume that the photographers were either consumed by patients with active ANSD, or were themselves infected and thus terminated in the incineration.

Initial images were suspected at first by some to be an elaborate hoax (7, 8). Indeed, the similarity to "zombie" scenes from popular culture of the first photographs and videos seemed more science fiction than real. However, epidemiologists and medical historians have since postulated that small outbreaks of less virulent cases of ANSD have occurred throughout history, especially in the largely low-pH environments of Central Africa and tropical South America, and that these cases potentially created the genesis for the zombie myth (9, 10). Although now banned by UN protocol, pirated copies of these initial images from the Aran Islands are still widely available at a number of illegal Internet sites and represent a major psychological public health concern (11).

Initial documentation suggests that Inis Oírr, the easternmost island among the Arans, was overrun on approximately 6 May 2011 by local inhabitants who were first presumed inebriated or somehow intoxicated.

Men, women, and children in quick succession began displaying the now familiar ataxia, massively decreased frontal lobe function, cognitive decline, and profound hunger. It is believed that the hunger coupled with the cognitive degeneration is what triggered the aggressive outbursts, and indeed current investigations suggest that the still-unknown infectious agent causes increased foraging and hunting behavior most consistent with reptilian neurobiology. This is what led to early characterizations of the disorder as Reptilian Aggressive Hunger Syndrome, or RAH (12, 13).

Countless stories of outbreaks around the world soon emerged. They all showed the same presentation: ataxic ambulation among humans of all ages, with dysregulated sleep–wake cycles, aggressive attacks on all forms of visible fauna, and special and still poorly understood preference for evisceration prior to feeding. Occasionally mammals and birds appear similarly infected, though the disease appears to prefer human hosts, and attempts at animal models have not been successfully replicated. Reptiles, amphibians, and all invertebrates except squid, octopi, and other cephalopods remain immune.

Three early reported survivors of ANSD attacks all noted relentless pursuit by infected individuals (14), but not the necessarily massively increased strength as was previously thought. These survivors also noted the apparent fatal vulnerability that ANSD humanoids display to head trauma, leading to directives in the current *UN Self-*

Defense Manual that call for security forces to strike ANSD humanoids in the cranial region (15). It was in fact this early observation that suggested the presence of increased cranial pressure among ANSD humanoids, something current laboratory investigations have now definitively verified (16, 17, 18). Additionally, heat-scanning satellites and remote devices sent into infected areas prior to incineration have documented increased body temperature, sometimes as high as 105°F, among infected humanoids, consistent again with reports of those who have survived initial attacks but had physical contact with infected individuals. However, large-scale attempts at artificially cooling infected regions or individual humanoids have proven futile and were abandoned three months ago. It is important to note as well that all known survivors of attacks by ANSD sufferers have eventually succumbed to symptoms of the disease within two to three days, and all who have studied either survivors of these attacks or ANSD humanoids themselves have until recently become infected (13). This observation led to the United Nations resolution of 10 July 2011 calling for immediate controlled incineration of all infected geographic regions (19).

ANSD quickly spread throughout the globe, with the greatest devastation in population centers. Paris, France; Christchurch, New Zealand; and Reykjavík, Iceland, were the first affected cities, and following initial nuclear incinerations little remains of these once vibrant populations. Epidemiological investigations now track three tourists who had visited the Aran Islands in mid-April 2011 as the main vectors for bringing the disease to these population centers. Further outbreaks throughout June 2011 led to increased and improved early surveillance, and the UN program of now non-nuclear incineration—primarily through the use of neutron radiation—has slowed the spread of infection substantially, but not without enormous and potentially irretrievable human and cultural costs, as well as substantially increased greenhouse gas emissions. In particular, the increased warming caused by the mass incinerations may be propagating the infection more efficiently and deserves further study (20, 21).

Using gossamer technology, the first infected individuals were quarantined in Geneva at the World Health Organization on 12 July 2011. These individuals were kept in sterile, level 4 holding environments approximately two hundred feet beneath the surface of the Earth. Although the infection reported in Old Town, Geneva, may have been incidental to the quarantined humanoids, the possibility that the infection may have breached the holding area in Geneva and the eventual necessary incineration of the city itself led to the establishment of the sanctuary on Bassas da India, a small uninhabited coral atoll between Madagascar and mainland Africa. This location was chosen for its proximity to land, allowing scientists to quickly arrive once those already on the island expired, and for its consistent tropical temperatures (22, 23). As noted, the infectious agent thrives in warmer temperatures, and quicker and more virile progression of disease is therefore possible at the sanctuary among study subjects. The world medical community responded with extraordinary proficiency in creating the facilities at the sanctuary, though understandable but now regrettable ethical debates clearly hindered early investigations (24).

To date, there have been more than two hundred million confirmed infections of ANSD. The infectious agent is not known and the method of disease transmission remains elusive. Rate of onset from time of exposure is also uncertain, though current estimates place disease onset between two hours and three days post-exposure (12, 13). Infectious likelihood appears highest in areas that have seen the largest proportional increase in temperatures from the last century (20). This had led some to implicate increased global temperatures in the proliferation of the illness, though this has yet to be definitively determined (20). Forty-three major cities, defined as population centers with greater than 150,000 residents, have been entirely incinerated, and most of humanity is under martial law to preserve orderly food distribution and infection surveillance.

Technology combining exogenous high-dose Lasix and the infusion of bicarbonate have allowed pathologically high pH blood-gas measurements among scientists to stave off infection while studying infected humanoids (25, 26, 27). However, conditions creating the necessary metabolic alkalosis inevitably lead to decreased cognitive functioning, cyanosis, and respiratory arrest (28). The first scientists to induce these abnormalities in acid–base balance died relatively quickly. Their work was invaluable in perfecting the technique that now allows for the longer time period investigators can survive before succumbing to the compensatory mechanisms that lead to death in response to the pH changes. To date, no investigators have studied patients with ANSD without expiring either from attacks and evisceration by study subjects, or from forced incineration after contraction of the disease itself, or secondary to the acid–base changes necessary to stave off infection in the first place. The average life span of ANSD investigators after arriving on Bassas da India is predicted to remain fixed at approximately seven weeks or less (26). Conditions are without question dire.

NEUROBIOLOGY OF ANSD

The extent to which patients with ANSD mimic the popular depiction of zombies in film and literature is startling. As in film versions, ANSD patients seem to lack even rudimentary intellectual functioning. Some early investigators attempted to use so-called zombie films as a means by which the natural history of ANSD might be best understood. In fact, early studies showed difficulty among emerging experts in distinguishing movie footage from actual media coverage of ANSD outbreaks (12). This led to the infamous censorship of all zombie fiction and movie material (29). Nevertheless, the early studies of patient and film validity directly spawned some of the initial and startlingly accurate neurobiological conclusions that current postmortem analyses have confirmed. Since that time, more has been learned about ANSD neurobiology, with significant pathological findings illuminated in the hypothalamus, basal ganglia, cerebellar functioning, higher cortical functioning, autonomic peripheral nervous function, and limbic apparati (30). Comparisons have been drawn to other contagions that cause hyperaggressive behavior and hunger via encephalopathic changes (31). Less certain are early hypotheses regarding the potentially more focused use of cortical activity in the enactment of pack behavior seen among wolves and wild dogs. Certainly some media footage has suggested that humanoids with ANSD hunt with cooperative efforts, something unlikely if the previously hypothesized cortical dysfunction were globally present (31). The endeavors on Bassas da India will undoubtedly yield important confirmations of these early biological hypotheses and hopefully yield new findings toward a global remedy. The future of humanity rests in these endeavors.

CONCLUSION

ANSD represents a significant threat to humanity. The infectious agent is unknown, it is highly contagious, and there are no known cures or vaccines. This manuscript is intended to serve as the basis for further intensive research and study. Boundary disputes, international conflicts, and other political disruptions must now yield to the work of the ANSD Working Group. There are no remaining options.

REFERENCES

1. Golan, E. et al. Epidemiologic Predictions of RAH—the Rapid Spread of the Zombie Plague. Special Edition: *Journal of the American Medical Association* (35), pp. 2896–2905, July 2011.

2. United Nations ANSD White Paper #219: Bassas da India as Secure Study Site for ANSD—a Global Priority. July 2011.

3. United Nations ANSD White Paper #301: The Ecumenical Treaty of Atlanta. July 2011.

4. ANSD Working Group on Disease Classification. International Classification of Disease—Criteria for Stages I–IV ANSD Infection. WHO Emergency Session. June 2011.

5. United Nations Working Group on Epidemiologic Modeling for ANSD: A Genuine Threat to Human Survival. August 2011.

6. O'Flannagan et al. Description of Zombiism in the Aran Islands—Hysteria or Cause for Alarm? Irish Medical Society Web site, May 2011. www.celticmedicalsociety.org.

7. Bertram, L. and Annapolopedes, M. The Cultural Context of Multi-Player RPGs: The Reach of Zombie Populism. *International Journal of Anthropology.* pp. 465–490. May 2011.

8. Cornstein, J. Don't Fear the Reaper: Laughing at Death Across the Internet. *Journal of Popular Culture and Cyberspace,* pp. 346–360, May 2011.

9. Wilkins, Harold T. *Secret Cities of Old South America.* Adventure Unlimited Press. Kempton, Ill. 1952, p. 65.

10. Davis, Wade. *Passage of Darkness: The Ethnobiology of the Haitian Zombie.* Robert F. Thompson, Richard E. Schultes. University of North Carolina Press. 1988.

11. Birnbaum, J. and Coone, L. PTSD Following Illegal Viewing of Aran Island Outbreak Footage in a Large Community Sample. *American Journal of Psychiatry,* July 2011, pp. 2456–2460.

12. Hunter, M. et al. Zombiism and the Reptilian Brain: Is RAH Neurobiologically Sound? *Nature,* July 2011, pp. 145–149.

13. Zimmerman, G. Reptilian Aggressive Hunger Syndrome: Proposed Terminology for Endemic Zombiism. *International Journal of Public Health,* July 2011. pp. 23–29.

14. Gorgon, M. et al. Not Strong, But Fast. Pursuit Behavior by ANSD Humanoids. *International Journal of Animal Behavior,* July 2011, pp. 27–35.

15. ANSD Public Health Advisory: Hit Them High—Recommended Attack Postures for ANSD Stage IV Humanoids. July 2011.

16. Gupta, A. and Murwazaki, M. Evidence for Increased Cranial Pressure in RAH. *Journal of Virology,* June 2011, pp. 345–360.

17. Gupta, A. and Johnson, S. Increased Cranial Pressure, Prions, and RAH—Etiology for Cranial Vulnerability. *International Journal of Neuroscience,* July 2011, pp. 68–70.

18. Kumar, G.; Kalita, J.; Misra, U. K. Raised Intracranial Pressure in Acute Viral Encephalitis. *Clinical Neurology & Neurosurgery,* June 2009, 111(5): 399–406.

19. United Nations Emergency Resolution. Controlled Incineration Through Nuclear and Non-Nuclear Means for ANSD Infected Areas with Populations Greater than 150,000, August 2011.

20. Gupta, A. and Blum, S. Does Heat Accelerate ANSD Virulence? Missteps from the UN Early Resolutions. *International Journal of Virology,* September 2011, pp. 20–25.

21. Garbe, T. R. Heat Shock Proteins and Infection: Interactions of Pathogen and Host. *Experientia.* 48(7): 635–639, July 1 1992.

22. http://www.nationmaster.com/country/bs-bassas -da-india.

23. UN/WHO/ANSD Working Group Press Release. Bassas da India to Be Location of New United Nations Sanctuary and Study Site for ANSD. (UNSaSS). August 2011.

24. O'Flynn, M.; Bernstein, L.; Hiawatha, M.; Steadle, N. et al. Reply to the Treaty of Atlanta. Life, No Matter How Wretched, Is More Sacred than Death. Objections to Current Practices with Regard to ANSD. *Journal of Medico-Ethics,* August 2011.

25. Supattapone, S. Prion Protein Conversion In Vitro. *Journal of Molecular Medicine,* 82(6): 348–356, June 2004.

26. Gutierrez et al. Artificial Metabolic Alkalosis Stops Neurologic Progression of ANSD. *Concepts in Virology,* August 2011, pp. 34–56.

27. Bohmig, G. A.; Schmaldienst, S.; Horl, W. H.; Mayer, G. Iatrogenic Hypercalcaemia, Hypokalaemia and Metabolic Alkalosis in a Lady with Vena Cava Thrombosis—Beware of Overzealous Diuretic Treatment. *Nephrology Dialysis Transplantation,* 782–784, March 1999.

28. http://io9.com/5286145/a-harvard-psychiatrist
-explains-zombie-neurobiology-addendum,
May 2011.

29. Johnson, M. and Blum, S. Controversy Regarding
Zombie Censorship. *New York Times,* June 11, 2011.
(Editorial)

30. Gupta, A.; Johnson, S.; Martinez, J.; Blum, S.
Early Neurobiological Postulates for ANSD. UN/
WHO/ANSD Working Group. *Nature-Neuroscience,*
July 2011, pp. 46–78.

31. Gorgon, M.; Johnson, S.; Gutierrez, B. Are Humanoids Evolving? Reports of Pack Behavior Among
Hunting ANSD Humanoids. *New England Journal
of Medicine.* Commentary, August 2011, pp. 57–60.

APPENDIX III

United Nations
General Assembly Distr.: General
18 July 2011
Sixty-third session
Agenda item 1
09-9876

EMERGENCY RESOLUTION ADOPTED BY THE GENERAL ASSEMBLY

WITH:

COOPERATION FROM THE WORLD INTERFAITH COUNCIL

PERTAINING TO:

STATUS OF HUMANS SUFFERING FROM ATAXIC NEURODEGENERATIVE SATIETY DEFICIENCY SYNDROME (ANSD)

The General Assembly,

Reaffirming the purposes and principles enshrined in the Charter of the United Nations and the Universal Declaration of Human Rights,[1] in particular the right to freedom of thought, conscience and religion,

---

1. Resolution 217 A (III).

Recalling its resolutions 56/6 of 9 November 2001, on the Global Agenda for Dialogue Among Civilizations, 57/6 of 4 November 2002, concerning the promotion of a culture of peace and non-violence, 57/337 of 3 July 2003, on the prevention of armed conflict, 58/128 of 19 December 2003, on the promotion of religious and cultural understanding, harmony and cooperation, 59/23 of 11 November 2004, on the promotion of inter-religious dialogue, 59/143 of 15 December 2004, on the International Decade for a Culture of Peace and Non-Violence for the Children of the World, 2001–2010, and 59/199 of 20 December 2004, **on the re-examination of these fundamental principles with regard to humans suffering from ANSD,**

Underlining the importance of promoting understanding, tolerance and friendship among human beings in all their diversity of religion, belief, culture and language, and recalling that all States have pledged themselves under the Charter to promote and encourage universal respect for and observance of human rights and fundamental freedoms for all, without distinction as to race, sex, language or religion,

Taking note of the adoption of the 2005 World Summit Outcome[2] in which the Heads of State and Government acknowledged the importance of respect and

2. See resolution 60/1.
   A/RES/61/2212

understanding for religious and cultural diversity, reaffirmed the value of the dialogue on interfaith cooperation and committed themselves to advancing human welfare, freedom and progress everywhere, as well as to encouraging and promoting tolerance, respect, dialogue and cooperation at the local, national, regional and international levels and among different cultures, civilizations and peoples in order to promote international peace and security,

Alarmed **that the nature of ANSD infection threatens <u>all people of this Earth equally and without prejudice,</u>**

Aware **that those suffering ANSD constitute <u>significant threats to the future of Humanity,</u>**

Emphasizing **that ANSD leaves its victims <u>no longer able to exercise the most basic human attributes</u> that qualify for protection and respect in accordance with international guidelines,**

Reaffirming **that basic human attributes are a necessary component to the rights and privileges afforded by all nations to all peoples,**

Affirming the need that all nations, cultures and ethnic groups view the ANSD Pandemic with identical, similar or complementary philosophical, ecumenical and existential alarm,

Considering that cultural, ethnic, religious and linguistic diversities are currently significantly at risk given the rapid spread of the ANSD Pandemic,

Recognizing the magnitude of this risk as noted by

the World Health Organization and other International Bodies in the absence of a current vaccine, cure or lasting treatment for ANSD,

Taking note of the valuable contribution of various initiatives at the national, regional and international levels, such as

- **The International ANSD Vaccination and Treatment Programs,**

- **The Bali Declaration on Building Interfaith Harmony Within the International Community,**

- **The ANSD/UN Working Group and Plague Eradication Collaborative,**

- **The Congress of Leaders of World and Traditional Religions Special Conference on Human Attributes,**

- **The Dialogue Among Civilizations and Cultures, Enlightened Moderation,**

- **The Informal Meeting of Leaders on Interfaith Dialogue and Cooperation for Peace,**

- **The Islam–Christianity Dialogue,**

- **The Second Moscow World Summit of Religious Leaders Convened to Discuss the Rights of ANSD Victims, and**

- <u>The Tripartite Forum on Interfaith Cooperation for Peace,</u>

which are all mutually inclusive, reinforcing and interrelated,

Mindful that the cooperative initiatives for practical action in all sectors and levels of society for the promotion of **ANSD TOTAL ERADICATION** and the subsequent **PERPETUATION OF HUMANKIND,**

Recognizing **the commitment of all religions to peace,**

1. *Affirms* that mutual understanding and inter-religious dialogue constitute important dimensions of an active and definitive world response to Ataxic Neurodegenerative Satiety Deficiency Syndrome, also known as ANSD, also known as zombiism;

2. *Takes note with appreciation* of the work of the United Nations Educational, Scientific and Cultural Organization on inter-religious dialogue in the context of its efforts to promote dialogue among civilizations, cultures and peoples, as well as activities related to a culture of peace, and welcomes its focus on concrete action at the global, regional and subregional levels and its flagship project on the promotion of interfaith dialogue;

3. *Recognizes* that respect for religious and cultural diversity in an increasingly globalizing world contributes

to international cooperation, promotes enhanced dialogue among religions, cultures and civilizations and helps to create an environment conducive to the exchange of human experience;

4. *Also recognizes* that, despite intolerance and conflicts that are creating a divide across countries and regions and constitute a growing threat to peaceful relations among nations, all cultures, religions and civilizations share a common set of universal values and can all contribute to the enrichment of humankind;

5. *Reaffirms* the solemn commitment of all States to fulfill their obligations to promote universal respect for, and observance and protection of, all human rights and fundamental freedoms for all in accordance with the Charter of the United Nations, the Universal Declaration of Human Rights and other instruments relating to human rights and international law; the universal nature of these rights and freedoms are beyond question;

6. *Urges* States, in compliance with their international obligations, to acknowledge that:

I. **Human Beings are from this point forward solely defined as possessing the capacity for:**

   a. **Social Connectedness (attribute "a"), unless absence of or deficiencies in attribute "a" are** the result of a medical condition or conditions other than Stage IV ANSD

b. **Emotional Recognition of Others as Sentient Beings (attribute "b"), unless absence of or deficiencies in attribute "b" are the result of a medical condition or conditions other than Stage IV ANSD**

II. **Furthermore, Human Beings afforded protection by all States:**

c. **Do not threaten via ANSD contagion and Stage IV infection the health of Humanity on both individual and global levels (attribute "c")**

d. **Are in all cases considered Human when attributes "a" and "b" are not the result of neurological, psychiatric or other medical conditions** in the absence **of Stage IV ANSD infection**

7. *As such*, it is acknowledged that:

I. **Humans infected with Stage IV ANSD— neurologic changes that promote a-cognitive aggression and cannibalism in the absence of sound recognition of the moral inconsistencies therein as a function of ANSD infection—are NOT HUMAN, and will be from this point forward referred to as:**

II. **Humanoid or**

III. **No Longer Human (NLH)**

8. *As such*, it is acknowledged that:

I. **The promotion and protection of the rights of persons does not apply to those with Stage IV ANSD**

II. **The protections afforded human and other biological subjects in the course of medical experimentations are not required for those with Stage IV ANSD**

III. **All citizens may with impunity destroy Stage IV ANSD Humanoids**

9. *With Regard to Medical Experiments*, it is acknowledged that:

I. **Great haste is necessary**

II. **Current neurobiological evidence suggests a lack of somatosensory input, rendering such actions as general anesthesia unnecessary and time consuming**

10. *With Regard to the Onset of Death in ANSD Humanoids*, it is acknowledged that:

I. **Death occurs at the time of ANSD Stage IV infection**

II. After Stage IV infection, Humanoids are considered "Animate but Not Living"

III. Appropriate funeral rites and other ceremonies marking the passage of life are permitted whether or not the ANSD Humanoid body is available

11. *Simultaneously, it is emphasized that:*

I. Needless acts of violence against Stage IV ANSD Humanoids are not ethically, religiously or socially sound

II. It is never ethically, religiously or socially sound to destroy a person who is suspected to have ANSD but lacks signs and symptoms of Stage IV infection

III. Those who engage in such acts are subject to prosecution as afforded by the rights of individual States

12. **After which time that a viable ANSD cure emerges,** these guidelines will require significant reinterpretation.

APPENDIX IV

This material was discovered on Blum's private laptop in his quarters at the UN bunkers. The dates suggest this was written in the days preceding his transfer to the island. We have included this material to accentuate the extent to which Blum was scientifically ill prepared to undertake the mission he ended up heading, but nevertheless entirely emotionally capable and willing. Tomorrow's agenda will undoubtedly include plans for future missions. Blum's psychological preparations are at least as important as whatever scientific knowledge he brought to the team. His mental and psychological state therefore represent key components of a successful mission. Please read this material with these observations in mind.

There will be a short memorial tomorrow morning in honor of Dr. Blum before the meeting commences.

November 10, 2012

I went up above today for the first time in months.

Is this our planet? Our home? Where *is* this place?

Everything is green, hazy. Windy, and the air smells.

I had forgotten about the smell, forgot to bring a mask. I retched a bit but recovered. Wiped my mouth and went for a walk.

You can only go so far on the bunker walls—there are humanoids scattered around the perimeter, all deanimated. Trees without leaves. Bare trunks sway in the wind.

The last human arrived about four months ago. He was floating by himself on a little dinghy; it was painted blue and orange, a lot like the sky used to be. All this gray surrounding his boat. I think we would have missed him if his boat hadn't been painted.

At that time, there were still humanoids on the island. They were starting to eat whatever moved, whatever was living. Odd how they continued to avoid one another.

The flags above this compound still fly—three flags, actually. There's Old Glory, one for the UN, and one for the WHO.

But there've been a lot of traps out there, too, people flying one or more of these flags to attract survivors, usually for food. This guy trusted us. Had to, I guess.

Good God, what have we become?

This guy, the one in the boat, sat there screaming, paddling with his hands, trying to direct his little boat over the gray dead ocean. Sensors confirmed that he was alive, febrile but still human. A droid picked him up and brought him in. He spoke some island language, not sure we placed it. Spoke a bit of French, too, and he could say thank you in English. Said it a lot. We fed him, gave him a bed, heard him cough, and gave him some Lasix. Got his pH up and his walking improved.

He changed five days later. Didn't take the Lasix and tore the pH meter out of his belly. Didn't trust us. Can't blame him. Hard to trust anyone healthy. We shot him and tossed his body over the wall. No one knew his religion, so we got the Unitarian folks to say a few words.

That's what it was like out here four months ago. That's our world.

That's the planet we're trying to save.

November 11, 2012

Why'd I go up there? Why'd I go for that walk? I already knew I'd volunteer. I know Blanca, knew Pittman a little. I already knew I would go.

The transmission was definitely Blanca's voice. I recognized it right away, even through the static, through the coughs. I knew way before we ran the voice scan.

She's practical, almost to a fault, and probably sick as hell. She wouldn't waste energy on a transmission unless she was on to something. She may be the smartest person on Earth right now. I spent some time combing through her old notes, trying to make sense of that message. Lots of work on prions, on the ways they turn evil with acid. Lots of theories on the way it was incorporated in influenza. She wasn't convinced it was an accident.

But nothing that would help me to make sense of her message. Just notes on influenza, prions, and memories of the Spanish countryside. She missed home. Wrote a lot about shepherds in Andalusia, compared them to the shepherds in Ireland, where all this started.

I had to decide, I guess. Had to decide whether it's all worth saving.

How could things go so wrong so fast?

I wonder about Sarah and Abe.

I know they're gone—hard to imagine them still around. Tried to warn Abe, but he said he had a congregation to lead, a responsibility. Then send me the kids, I said. Let them take cover, but his wife refused. There are other worlds, Dad. That's what he said.

I wonder what world he's in.

We've heard almost nothing from the West Coast. The nukes triggered the faults. Big ones—8.5 and higher. Can't imagine Sarah made it.

I'm glad Miriam was gone before all this started.

But this place isn't life, either. Concrete walls, barbed wire. Sometimes the wind is so loud you can hear it even four stories down. You can only watch so many movies, read so many books. There are cockroaches. Really big ones. I guess they have lots to eat.

This is America. A fortified outpost in the South Pacific. They say there are still humans on the mainland, but we've stopped our rescue efforts. The Chicago exodus was a nightmare. Helicopters leaving, people hanging on to landing gear. All those families on the tops of buildings. We left clergy behind and some cyanide.

Where would they go? All the islands are overrun, humans living like animals. Africa, Europe, Asia—all barren, wastelands. Antarctica? We can barely keep the water clean here.

But the Crypt also looks like hell. The temperature averages ninety degrees, the humidity constant. Everyone dies there, one way or another.

But I can't stay here. I can't stand the fluorescent light anymore. I hate the bunks. The mattresses smell like bleach. And I hate the whole military thing. How can I be a captain? Captain of *what*? People salute me in the hall and I forget to salute back. Not even sure I know how to salute. They stand there waiting, like it still matters. We don't even have a brig. That's what outside the walls is for.

I am Stanley Blum. *Captain* Stanley Blum, division leader of the Administrative Corps of the ANSD Working Group.

And I can't even tell my mother I got promoted.

I've given this lots of thought and I want people to know.

I volunteer. I'll go. I know I won't come back, but I'll go. Why in *this* world wouldn't I?

APPENDIX V

The following documents were found on the hard drive of a personal computer in the air lock of a badly damaged yacht moored off the coast of southwest Greenland. The ship was deserted, with no signs of recent life or humanoids aboard, and these documents are the printed remains of all that was salvageable from a highly encrypted set of communications. We have included them here as they came to the attention of our investigators after we began searching all encrypted digital communications for terms that might suggest classified communications regarding ANSD. So far, there are over four thousand documents and we will continue the process of screening this material before bringing them to our working group for analysis. We have no knowledge of any of the names or possible code names below. As Blum had concluded that the ANSD pathogen is manufactured, we will continue to follow all leads.

To: 2011@mobius.net
From: KFC
Subject: movement?
Agreed.

Markets will initially fail.

Our failure to capitalize on the last bubble was a colossal fuckup, and my ass still hurts from where Adrian chewed me a new one. Shit, we created the bubble—we could have done better.

But this is really big. Lots of moving parts. I'm just not comfortable making the call until I get approval from The Gate. Where the fuck has he been. Apparently even Adrian doesn't know.

Hope you and the family are well. Thanks for the birthday photos. Was that nanny dating anyone? I'm just asking. Sharon's getting pretty, by the way—you better keep an eye on her :)

Listen. Get back to me.

K

* * *

To: Adrian
From: 2011@mobius.net
Subject: KFC
A—

Heard from K. He's got cold feet, I guess, or there's something you're not telling me. I gotta manage him, so can u let me know whether this thing has gotten more complicated? Hasn't the vaccine production started? We had 100% response with the baboons, didn't we? If we did, then what's K all twitchy about? He keeps talking about moving parts.

I ran the projections again, this time adjusting for 50% resistance. I ran them for airborne and water vectors.

Adrian, we can't miss. Market'll plunge—no flights, no work, nothing really for the first month or so. That's when we place our options. I figure, with the hedges we've grown, those puts should yield 200, maybe 400 million in the first week.

Then, with the start-ups and ancillary investments, an additional billion or more.

Talk to the Gate. This looks good for us. I don't want to miss out.

Me

*   *   *

To: 2011@mobius.net
From: Adrian
Subject: re: KFC
Jesus, I got grandkids now. I'm done. Leave the Gate out. He's crazy, talking like one of those televangelists. Let's ride the market through. The company has soft enough cushions.

Hey, I shot 4 under yesterday. 4 under! Couldn't do that with a fever and cough, now, could I?

Relax. Spend more time with your kids. They grow up fast.

My love to your mom.

A

*   *   *

To: KFC
From: 2011@mobius.net
Subject: re: movement?

Adrian's out for now, but that doesn't mean we should stop. We're all VPs, and no one made him king. What're you worried about? What could go wrong?

I'll try to find the Gate. I think I read somewhere that he was shark fishing in the North Sea... Can you believe that? Talk about irony! I guess he can think like them. Adrian says he sounds a little nutty these days, but I want to talk to him directly. I'll get back to you as soon as I reach him. Don't make any calls until then. We don't want this out till we're ready.

And, yes, Sharon's a cutie. Doesn't get it from her old man, but she takes after Gloria, I guess.

Oh, and the nanny has a temper, but you

should see her on our cameras when she thinks she's alone in the house. There are perks to being a family man...

Anyway, I'll get back to you as soon as I know more. Don't worry about Adrian. He'll change his mind once he gets home from visiting his grandkids. He always does.

Go on a date or something. Get laid. It'll take your mind off all this money.

Me

* * *

To: 2011@mobius.net
From: KFC
Subject: re: re: movement?
I don't give a damn about the money anymore. I saw the lab, ran some more projections. Too many goddamn variables.

M, don't do this thing. Think of Sharon.

K

APPENDIX VI

GUTIERREZ'S PRIVATE JOURNAL (TRANSLATED FROM SPANISH—A LETTER TO THE PASTOR OF HER CHILDHOOD PARISH)

Dearest Father Benito:

I continue to struggle with the "NLH" distinction. It feels legalistic and rationalizing; it is not at all grounded in science OR religion. It is about moving forward, having some kind of plan, some response to the outbreak, except it really feels like moving backward. Things were barbaric enough before this plague. At least until now we've had our humanity. Now we have more narrowly defined even that.

Is someone born with a genetic malformation human? Their genes are different, and genes, or at least chromosomal counts, are how we've always defined species. But someone with abnormal chromosomal counts is still human. Those born without limbs, without speech, without the capacity to feel...someone born even without a brain is still one of us. You taught me that. We are all God's creatures, all created in His image.

I have just accepted a position at the newly established UN bunkers in the South Pacific. This is a secret compound, but I am tired of secrets. I do not think this is premature for me to leave if I am to do any good, and if there were any way I could take you and all of the children of the village with me, I would. But then I couldn't go, because they wouldn't let me. And I'd stay here, along with you and everyone else, and we'd waste away, and change, and we'd feed. Would we still be human, Father? Would you? If we changed, would we still be able to receive His blessing?

So, I ask your pardon.

This note is like Confession. I can remember you when I was a child. You chuckled, and don't think I didn't hear you! You chuckled when I admitted to taking some candy from my brother. A "Hail Mary," you gave me, and then you wished my family well and me. Isn't that against the rules, to let me know so directly that you know me?

And you kissed me on both cheeks when I came back to our village after completing my studies. You praised my commitment to God's miracles of life. You told me I would save many lives, and you told me you were proud. When Papa died, you said that also. How did you know, when I was so young, that I would always want to know more?

And now, I have had to make a choice that only the Devil himself could devise. I have created a protocol, a method, to stave off the disease, to slow the progression of the horrible madness, but in so doing I doom us all to die as humans, make martyrs of those who do nothing but breathe.

I have never felt this way before. I have always known of bad answers. But what to do when the questions themselves no longer suffice?

GUTIERREZ'S INITIAL DESCRIPTION OF ACID–BASE MANIPULATION TO DECREASE VIRULENCE OF THE PRION COMPONENT OF ANSD

A. Primary Known Facts:

1. There is a prion component to ANSD.

2. Prions in ANSD infect higher brain regions first.

3. Progression of prion infiltration into brain tissue increases stages of ANSD (i.e., increased prion

infection within an individual brain causes progression from Stage II to Stage III and so forth...).

4. Prions are more virulent and damaging in low pH—more acidic—environments.

5. This is because prions bend into more actively infectious shapes with decreasing pH.

6. Infection from ANSD increases the normal immune response of the body, resulting in a lowering of pH—an increase in acidity—thus propelling the prion infection.

B. Primary Hypothesis:

1. Can internal pH be artificially raised to stop prion infectivity?

   a. known fact—we CAN artificially raise pH

      i. administer bicarbonate

      ii. increase urination using medications

      iii. this process is itself potentially toxic

   • Toxicity from increased pH—seizures

   • Toxicity from increased pH—tetanus

   • Toxicity from increased pH—respiratory difficulties

- Toxicity from increased pH—confusion

- Toxicity from increased pH—dangerous heart rhythms

- Toxicity from increased pH—DEATH

  b. known fact—we CAN reverse the effects of artificially raising the pH

    i. administer acidic compounds

    ii. alter respiratory status

    iii. stop numbers i, ii, and iii from above

  c. known fact—reversal of artificially raised pH will lead to progression of prion infection

C. Primary Confounders:

1. Death can result from artificially increased pH.

2. NLH status ("death") can result from ongoing prion infection.

3. The symptoms of artificially increased pH are very similar to the symptoms of ANSD progression.

D. Necessary Components:

1. Must be able to measure internal pH in real time.

2. Must be able to alter internal pH in real time.

3. Must be able to monitor for abnormal heart rhythms in real time.

NOTE: The process will inevitably lead to "death" either via uncontrolled ANSD progression or fatal pH elevation.

APPENDIX VII

FLOW CHART FOR IMPLEMENTATION AND POSSIBLE OUTCOMES FOR TREATMENT OF ANSD

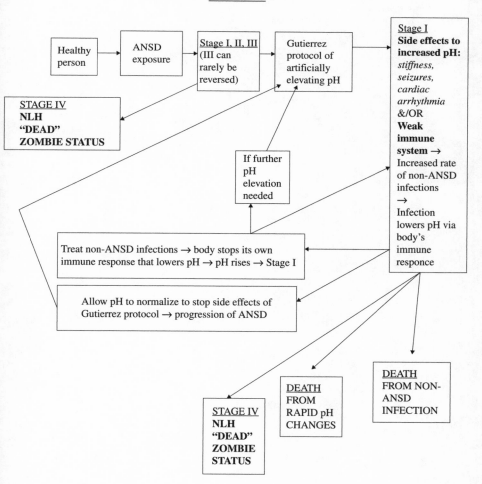

THERE HAVE BEEN QUESTIONS...

Here's the thing. If you tell people that you're going to write a Zombie book, that you have a major publisher like Grand Central Publishing willing to publish your Zombie book, that you will be attending—along with your usual set of work venues—gatherings with titles like "Scarefest" and "Spooky Empire," and that all the while you intend to continue in your position as a somewhat respected (or at least competent) physician at a relatively well-known and august institution (like Harvard), you sort of shut the conversation down. It goes like this:

Mentor/Boss/Spouse/Colleague/Parent and so **forth, hereafter referred to as**

"Anyone": *Hey Steve, whatcha doing these days?*

Me: *Well, I'm working on a Zombie novel.*

Anyone: *Uh–huh.*

(Mild smile from the Anyone who is addressing me. Steve, after all, has been working hard.)
Then there is **The Pause.**
Anyone tries again.

Anyone: *So . . . seriously, whatcha been up to?*

(Notice the feigned appearance of recognition that we are in fact just "shootin' the shit." But I'm a shrink. Subtle cues in language, changes in body posture . . . all this stuff *matters* to me. And Anyone isn't being subtle here. Anyone in fact looks pretty worried about me. But I'm having fun, some of the most fun I've ever had.)
So I tell Anyone: *I'm actually, really writing a Zombie novel.*

There is, of course, another pause. Then there are the inquiries, the concerns for my health, for the health of my family, for whether or not I am eating well or getting enough sleep. But, inevitably, Anyone just can't take it anymore.

Anyone: *Really? A Zombie novel. You mean like in the movies? Like the ones that eat people?*

Me: Yep . . . an honest-to-goodness Zombie novel.

Some of the Anyones then gather their strength and ask the really tough question.

Anyone: *Are they real? Zombies, I mean? They don't, you know, exist...do they?*

And that is how it starts.

This is the beginning of the very long list of inquiries, ponderings, curious concerns—all quite reasonable given the circumstances.

I've kept track of these questions (I have a file on my desktop called "Zombie questions"), and I have saved pretty much every e-mail I've ever received about the creation of this book. For this paperback edition, I'd love to answer some of these questions to set the record straight.

Let's start with the question that poor Anyone asks above.

QUESTION #1:

Are Zombies real?

I've learned to answer this question with the expectation that I will disappoint many people. Nevertheless, I respond truthfully, in part because I think I have an obligation to be truthful based on the oath I took as a physician, and in part because I am in general a crappy liar. Zombies, whether slow or fast, infected or contaminated, reanimated or never dead in the first place—ALL iterations of the cinematic wanderers who mindlessly crave human flesh—well, they ain't real. *They don't exist.*

Yet.

"Booooo!!" That's what I got at Comic-Con. "Hissssss," they said. They were bummed. These folks, lots of folks, wanted Zombies to be real, wanted the Walking Dead to be waiting for them at Starbucks or in the classroom or just standing there in an otherwise serene riverbed.

"Are you people crazy?" another of the panelists retorted. "Have you not seen the movies I've seen? A Zombie outbreak would totally suck."

But I get it, I think, or at least some of it. And I am always quick, therefore, to add that Zombies as we know them from guys like George Romero are brilliantly conceived ghouls that play to our worst and most iconic nightmares. Who hasn't been frightened in the dream when the attacker is slow but never stops? Who hasn't wanted to just end the madness of the pursuit of that slow attacker and blow its head off? And yet, there are always more attackers just around the corner.

Zombies are the feeling we get when we see a line out the door at the Department of Motor Vehicles. They're the ten billion other people on hold with you when you wait on the phone to talk to your HMO about why your bill is so high. They are the ones who are maddeningly unmoved, or at least appear to be maddeningly unmoved, by the pervasive stench of Muzak in elevators and call centers. They neither hate nor love *The Jersey Shore.* They're as mindless as the common cold virus that creeps back into your house at the start of every school year.

And in that sense, they exist, all right. *Because Zombie stories aren't about Zombies.* They're about people's reactions to Zombies. And I know that my desire to declare Armageddon appears whenever I am told to stand in that bedraggled queue that the bored-looking airline employees direct me toward whenever the airline has lost my luggage. (But remember: It's not about me. Hell, they lose *everybody's* luggage). And if I decide to go all Hulk on the airline, glowing green and muscular with radioactive rage, it really won't help at all. Raising any manner of Hell ain't gonna move that line. My tantrum will get me at best a dirty look and at worst a pair of handcuffs.

We can't all be Michael Douglas in *Falling Down*.

But, we could be Michael Douglas in *Falling Down* if there were Zombies.

So, just as Freud described dreams as forbidden wishes, perhaps the nightmare of the Zombie is the hidden verboten desire for societal permission to **Totally. Go. Berserk**.

Do Zombies exist? No. Do we act like they do? All the time.

And therein lies the fun and the wisdom and the abiding caution of the Zombie trope. We don't need Zombies as permission for violence. We need Zombies to get us to appreciate just how violent we can be.

This question passes, then, and I become aware that

I have once again delivered a lecture where a sentence or two would have worked just fine (you'd understand this tendency if you grew up in my family). And we make our way to the next most common question.

QUESTION #2

"Could we make one? Could we make a real Zombie?"

And that question has come to me from physicians and teachers and screenwriters and authors and my own parents, bless their hearts.

Popular Science magazine asked me if we could make a Zombie. NPR asked me this on both *Science Friday* and *On Point*. *The Boston Globe*, *The Boston Herald*, and *The Boston Phoenix* asked. Hell, the *Harvard Medical School Magazine* wondered about this, and the *Stanford Magazine* made it a focal part of their article about my book. To be sure, no one wanted to know *how* to make a Zombie. That would be like telling the enemy in a war how to build something really awful. They just wanted to know if it could be done.

And, again, just as with the inquiry about the existence of Zombies, I'd say, "not really." It is really, really hard to come up with a contagion or a brain lesion or a disease process that would produce the famous Hare Krishna

Zombie in *Dawn of the Dead*. Weirdly, the behavior of the Zombie isn't the hard part. There are lots of things that partially make people act like a modern horror Zombie. And I'm not talking about the more conventional and truthful Zombie of island and African religions. I'm talking staggering, flesh-eating, stupid as hell Zombies right out of the movies. We could get a human to look an awful lot like a Zombie, but that's not the hard part.

The hard part? *We can't raise the dead.* A true Zombie— and this comes to me from all sorts of reputable sources, including Mr. Romero himself—must first be dead and then come back to life. They must reanimate. And this just doesn't happen all that often.

So, as anyone who has read this book knows, I decided that "death" was itself a construct. This idea is not at all outlandish. If you work in a hospital, then you've very likely openly and seriously struggled with this issue. *When is the quality of life, the state of life itself, the same as being "as good as dead"?* I'm not taking a stand on the answer here, largely because I don't think there is a universal stand to be taken. These questions are uniquely and immensely individual inquiries. Culture, law, religion, science, and even existential philosophy all play a role. But, just as in the hospital, at a certain point we have to decide. For every case where this question emerges, we must find some consensus as a society. Otherwise, some argue, we mire ourselves in intolerable ambiguity.

If Zombies were out there and if every day we had more Zombies to fight, then we'd need to feel comfortable killing. As a rule, I don't believe that most humans are entirely comfortable killing anyone.

But we also know from our shameful human history that once we legalize our classifications to include categories that call others something less than human, we get better and more comfortable with our nasty sides. So my guess, and it is I think an informed guess, is that we'd decide that Zombies do not live at all and therefore there is nothing to kill. We de-animate Zombies. We don't annihilate them. Why?

First of all, we'd have pattern recognition. My book takes place in a world where Zombie movies are known and actually banned. We'd see those who are infected with ANSD and decide that they're Zombies, and then we'd decide from that observation that Zombies are dead, which leads to the belief that we aren't in fact killing anything at all when we bash in their heads as Shaun does with his cricket bat.

However, we have our pesky conscience to deal with. Neurobiologically, that conscience is some kind of epic battle fought with mirror neurons, frontal lobes, memories, and our primitive and competing desires to create and to destroy. If we take that internal battle away from our enemies (Zombies don't have a conscience; Zombies don't really have much of a brain at all), then we find our conscience resting substantially easier as we smash in a Zombie's head with a fireplace tool.

But look at what happens to Pittman in this book. Look at how he screws it all up. It ain't the Zombies that spread blood all over Gutierrez's papers. It's a still-functional human who brings us all down.

When it comes to Zombies, I'd say the biggest threat is ourselves. Give a man a gun and a slow-shambling Zombie, give that man a fifty-minute head start, and put that man near a bunker with food for a year. That man, nine times out of ten, is gonna shoot that Zombie. He doesn't need to, but it's a freebie. A one-way invitation to the dark side of all of us, and that lesson feels more real to me than any Zombie ever can be.

QUESTION #3

Are you more worried about the slow or the fast moving Zombies?

This came up for the first time at Spooky Empire, a wonderful annual gathering of horror enthusiasts in Orlando. I suggested there that we stage an Oxford-Style Debate in which I would resolve that "slow-moving Zombies are way more frightening than fast ones."

My opponent in the debate was formidable. His

qualifications were scary. Kevin A. Ranson, a well-known author and horror movie critic, was without reservation in his assertion that fast Zombies are way worse than slow ones.

And he might be right for some people. I think I lost the debate, though it was close judging from the response in the audience.

But I can't escape my nightmare, just as I can't escape a shambling corpse. A fast-moving Zombie has its sights set on me. On ME! It wants me, and that's at least worth something, right? I mean, if it's gonna tear out my viscera, I want it to specifically want MY viscera. At least then it's personal. And if its personal I'm gonna have a much easier time blowing its head off.

But the slow ones? The ones that stare at me with dead eyes? The ones that would just as easily kill a complete stranger as me? The ones that find me entirely unremarkable but will always turn towards me if I so much as gasp? They scare the pants off me precisely because it ain't personal. The slow Zombies are viruses incarnate, and viruses do not have preferences. That lack of awareness in the Zombie, that total blankness of life, gives me time to think. And time to think, for humans, is our best and worst enemy. In my book, time to think leads us to nuking half the planet. In my book, time to think leads folks to not believe a word about the spreading contagion until it's too late. And in my nightmares, time to think buys me nothing but the chance to screw up. And if I screw up, God help me.

I'll take a fast Zombie over a slow one any day of the week.

Thanks for reading the book.

ACKNOWLEDGMENTS

The idea for this book, like a good zombie movie I guess, came from the wonderful capacity we humans have to shake our heads and laugh at ourselves. In fact, like some kind of unexpected mutation in a not-yet infectious agent, the book kind of germinated, a blip on an epidemiologist's chart. Will this one spread? Will it go airborne?

Plus, an epidemiologist tries not to leave anyone out.

So, I must start by thanking the Coolidge Corner Theatre in Brookline, Massachusetts. Their Science on Screen series gave me an opportunity to put science and zombies on the same slide, and their sponsors—the Boston Museum of Science, the Boston Phoenix, and others—were willing to laugh just enough to write about the whole thing.

My students at all levels—medical students, residents, fellows—endured all sorts of conjectures, and whether out of respect or something less noble, they smiled and at least seemed amused at my blithering. Certainly no one ever disagreed. My colleagues at Massachusetts General Hospital and Harvard were wonderfully supportive. I think doctors would love to play a bit more...Jay Fishman, a doctor's doctor and an amazing infectious disease expert, eagerly volunteered his expertise. My mother-in-law, Dr. Mirdza Neiders, offered both loving support as well as spot-on accuracy with regard to proper histological protocol. Larry Kutner caught the bug and offered unrivaled mentorship and direction, passing the zombie germ onto my agent, Laurie Liss.

Laurie's guidance, wit, and business sense were rivaled only by her willingness to supportively laugh. The bug went airborne somewhere in New York, when Celia Johnson at Grand Central Publishing showed the first signs of disease and decided that there was value in charting the progression of the undead. She has been on target throughout—keeping the story infectious and accelerated. The plague was spreading.

Gene Beresin, my friend and mentor, showed early symptoms, and I'd guess he had clear disease early last fall. John Herman, Mike Jellinek, and Greg Stone soon followed. In hindsight, we were approaching pandemic. When I expressed the requisite skepticism, James Frosch assured me that this was all very possible. At this point, the infection was real.

George Romero, Max Brooks, and Michele Kholos helped me to find solace and relief in the realization that someone could catch the zombie bug and still be a decent and respected partner, father, and friend. Robert

Weinberg helped me to make sense of the legal rights of zombies. He is hard at work on a zombie bill of rights as we speak.

And what would any good disease story be without passing the bug itself onto one's family. My parents and my sister embraced the living dead with filial enthusiasm and amusement. There they were, spending their retirement and leisure watching zombie flicks so they could understand what the hell their son and brother was doing. My wife treated herself to her first horror film on the big screen, and she managed to enjoy herself. This from a woman with whom there have been many heated Netflix disagreements...true love. My kids watched *Scooby-Doo* and *Zombie Island,* and my little one giggled beautifully as she stumbled about the house telling us that she was a "bombie." The dog even got into it.

Finally, to all the zombie enthusiasts who went out of their way to get infected with this new strain of excitement. The Zombie Research Society, led by Matt Mgok, Rick Loverd at the Science and Entertainment Exchange of the National Academy of Science, and a bunch of nice kids dressed like zombies I met in Toronto who all said something like "this is way cool" when I got excited about the book. Here's hoping that those who read this can chuckle wisely at all the gore. We must always ask ourselves where we draw our lines.

ABOUT THE AUTHOR

Steven C. Schlozman, MD, studied English and biology in Northern California, and taught high school English and science before starting medical school in New England. After training in psychiatry and child psychiatry, he joined the faculty at Massachusetts General Hospital and at Harvard Medical School, where he is currently an assistant professor of psychiatry. A longtime fan of popular culture, and especially of horror films, he has written about movies, books, pop songs, and sports in blogs for the *Boston Globe, Psychology Today,* and in academic journals. Most important, he has wanted to write a novel for a very long time and, perhaps even longer, to write the biography for the back cover. He lives in suburban Boston with his wife and two daughters, a big and ill-defined dog named Corduroy, a very fat cat named Daisy, and a skinny black-and-white cat named Oreo. In retrospect, the inspiration for this novel probably derives from the unlikely longevity of his daughters' pet crayfish, who despite a purported life span of less than two years continues to live happily and for much longer in its tepid tank, feeding on lettuce and the occasional wisp of sliced turkey. It is the first true zombie that Schlozman has encountered, and looks only slightly healthier than the typical walking dead. This is Schlozman's first novel.